CW00880275

ZERO-WASTE GARDENING

Maximize space and taste with minimal waste

BEN RASKIN

Illustrated by Alice Pattullo

First published in 2021 by Frances Lincoln Publishing
an imprint of The Quarto Group.
The Old Brewery, 6 Blundell Street
London, N7 9BH,
United Kingdom
T (0)20 7700 6700
www.QuartoKnows.com

A catalogue record for this book is available from the British Library.

ISBN 978-0-7112-6233-1

Ebook ISBN 978-0-7112-6234-8

10 9 8 7 6 5 4 3 2 1

Design by Tina Smith Hobson
Illustrations by Alice Pattullo

Printed in China

Ben Raskin (Wiltshire, UK) is the Head of Horticulture and Agroforestry at the Soil Association in the UK, and a father of two. Ben got the gardening bug working on an organic vineyard in northern Italy and has worked in horticulture for more than 25 years, including a stint as Assistant Head Gardener at the UK charity Garden Organic. He is now often found planting trees.

*Thanks to Ruth for your endless patience and support.
all the Rossers, and especially Helen and Hamish for your skill in creating writing
time for me, and of course, to Monica for your ongoing belief in me.*

ZERO-WASTE GARDENING

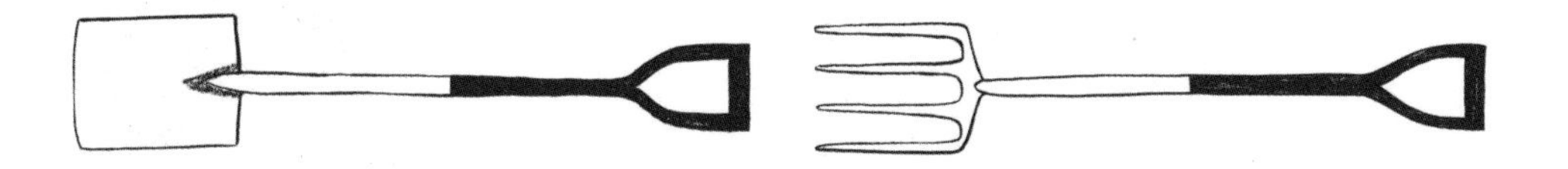

CONTENTS

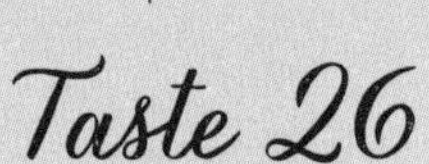

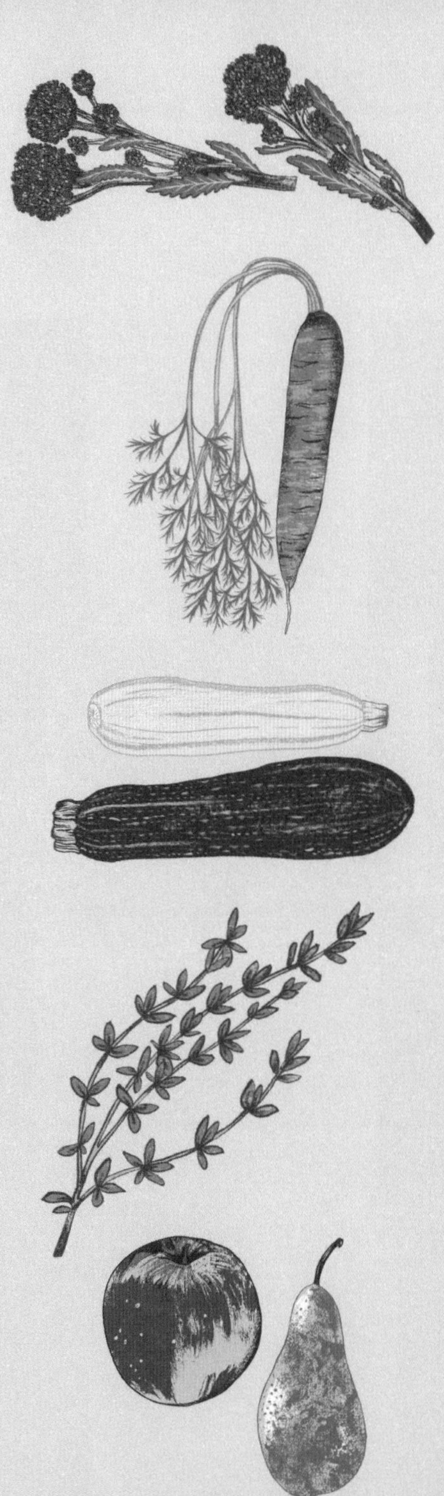

Introduction

THE PRINCIPLES OF ZERO-WASTE GARDENING

Zero waste is a popular term at the moment as we all finally wake up to the fact that our planet does not have infinite resources. Whether we can truly live in a circular sustainable cycle of resource use with our current population is still a matter of debate, but what is clear however is that we can all do more to help reduce our impact on climate and nature.

I am not going to pretend that you will become self-sufficient in fruit and vegetables, or will reduce your carbon footprint to zero. You'd probably need to work full time on about 2 hectares/5 acres to achieve that. However you can produce great tasting fruit and veg to supplement your bought-in food, and enjoy growing and cooking with produce from your own effort. Once we realize just how much of a plant we can eat and currently don't, we can perhaps make better use of the produce we buy as well as those we grow ourselves.

Where does waste happen?

BEFORE HARVEST

Not usually accounted for in food-waste figures, seeds that don't germinate are still waste. We've bought the seeds, spent time and effort sowing them, and probably watered them. All stages in the growing process have opportunities to lose a crop; pests, disease, drought, flood or even competition from weeds might all lead to crop failure. If we can improve our raising and growing skills we can reduce waste.

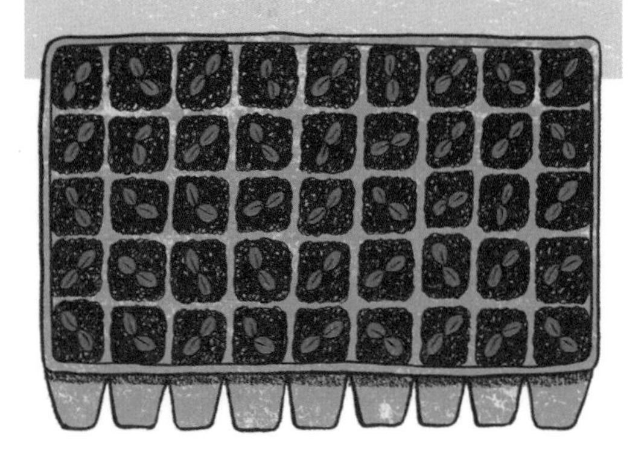

AT HARVEST

In commercial production, crop that does not meet the buyer's required specifications is 'graded out' at harvest. Produce can be graded out on appearance, size or shape. How much waste and what the specification is depend on who is buying it. Local box schemes will have more flexibility on shape and size for instance than supermarket buyers. When growing our own we can reduce that waste to almost nothing. If I've spent the time and effort growing it, I am going to use all I can, whatever it looks like. Using bits of the plant like carrot tops or vine leaves that you might not have thought of eating is another way to get more value and less waste from your crops.

IN STORE

Most produce needs to be stored for some period of time, sometimes just for a few days or, for crops like potatoes or apples, for many months. During that time some food will spoil. Improved storage conditions can reduce waste. This is where freezing, drying and pickling come into their own, drastically increasing the time we can safely store our harvest for.

IN THE KITCHEN

Eating everything you cook and monitoring what is in your fridge are good ways to reduce waste. My parents had lived through rationing in and after the Second World War, and leaving food on my plate was a definite no-no. With the planet in its current state, this is a sensible attitude.

OTHER INPUTS

As well as the actual crops, are there other ways we can become zero waste in our garden? Using recycled materials to make raised beds or trellises, using hand tools rather than petrol-driven ones, and maximizing renewable inputs like sunshine, water, composts and manures, are all ways to minimize the impact of our gardening efforts.

Space, taste and waste

Most of us have limited space for growing food. Choosing what to plant and making the best use of it once we have grown it can be tricky. How big do plants get? How much food do you harvest from each plant? What tastes better if you grow it yourself? We explore these questions, and also which bits of plants we can eat but don't., often even throwing them away. From carrot tops to cucumber flowers, fennel seeds to raspberry leaves, we will look at neglected foods to help you get the most from even a small plot.

SPACE

Zero-waste gardening means good planning; for each feature crop, there will be a guide to how much space one plant takes up and the yield you can expect. This will help you grow as much as possible of the crops you like, in the space you have. We'll also look at interplanting, where small quick crops can be squeezed between slower-growing ones, and at what plants grow up and be trained on walls or trellis to maximize your space.

TASTE

One of the great benefits of growing your own food is that it often tastes better than bought food. Which are the crops that really deliver flavour when you grow your own? Leafy vegetables are a must as they deteriorate so quickly once picked. Tomatoes, corn and peas are just a few of the others that are worth finding space for in your zero-waste garden. We'll also explore the different ways you can store and preserve crops for maximum quality and flavour. Freezing, drying, pickling and fermenting can help hang on to those gluts until you are ready to eat them.

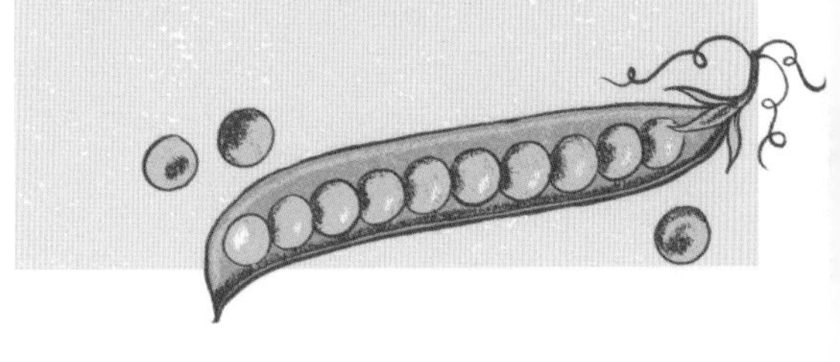

WASTE

It is important to grow what you need and then pick and eat it. If you go away in the summer, don't grow crops that fruit only while you are on holiday. What is the best crop spacing to grow the right size plant without leaving gaps? We'll also how to get free plant food, capture maximum energy from the sun, and gather nutrients from waste sources like coffee grounds.

Chapter one

SPACE

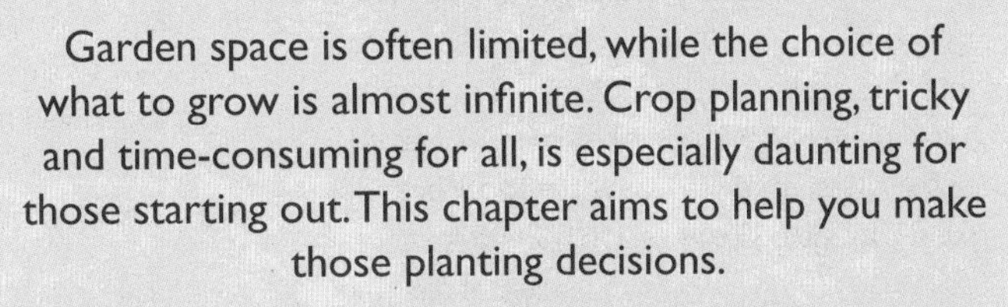

Garden space is often limited, while the choice of what to grow is almost infinite. Crop planning, tricky and time-consuming for all, is especially daunting for those starting out. This chapter aims to help you make those planting decisions.

Filling every available space can work in the short term, but we also need to think long term. Good crops need healthy soil, and good ground preparation and an understanding of crop rotations can help with this. A zero-waste garden is about minimizing inputs (fertility from outside the garden; costs; effort) and maximizing outputs by capturing all the sunlight and moisture we can. And we aim to use as much produce as possible.

Refer to the yield pages to see what might affect how much you actually get. As well as your own skill and experience, weather, soil and variety can all impact yields, as can pests and diseases.

Planning is key: successional sowing, training vertically to grow up as well as out Interplanting and undersowing are worth doing to gain space and minimize weeds. The 'How to Choose' pages give ideas on prioritizing which plants to grow.

Keeping records of what does well and when, is a great way to learn. Each year can be very different, and taking notes will help you to plan in future years.

PREPARING THE GROUND

Minimizing inputs/recycling

If you read the old gardening books on preparing ground you will find an obsession with digging and in particular with double digging, a strange concept dreamt up it seems by men who liked hard physical labour. I prefer to harness the power of nature to do my work for me, and we can incorporate this into a zero-waste approach to gardening.

When we disturb the soil by digging or cultivating we do two things. Firstly as we expose the carbon in the soil to the air it oxidizes, turning to carbon dioxide, which is released into the atmosphere. This is bad for our soil as we want carbon to feed soil life, and bad for the planet as we need to reduce carbon emissions. Secondly, cultivating not only disturbs the biological balance in the soil but also kills beneficial fungi. Fungi are particularly susceptible to disturbance as their hyphae (the white, root-like parts that spread underground) are so fragile.

For some crops, like carrots and parsnips, we don't have a choice. They need a fine soil tilth to be able to germinate well. For most plants though we can use a no-dig or reduced-dig system of growing that will lessen this carbon waste from our soil.

When we are taking on a new area to grow fruit and vegetables we need to remove or control the plants already growing there. An alternative to digging them out is to use a mulch – this is any material that covers the ground and stops light getting to the plants. This either kills them, or at least makes it easier to pull them out.

You can use layers of plain cardboard for this job. Remove the tape and lay the cardboard flat over the area you want to prepare for planting. It is a good idea to cut any existing plants as short as possible first. I usually put two or three layers of cardboard down to ensure a good barrier. You'll also need to make sure it doesn't blow away. You can just weigh it down with stones or something else heavy, but an even better way is to add another layer of organic material on top, like compost or woodchip. This not only keeps the cardboard in place but adds more carbon-rich food to your soil, which will benefit the plants when you put them in. If possible, lay the mulch in the autumn before a spring planting, but even a couple of months under cardboard will make the ground easier to prepare.

For fruit trees or large plants, we can leave the mulch in place, making holes to plant into. Even squashes can work like this if we don't have too many slugs. Most smaller crops perform better if we remove what is left of the cardboard and lightly fork over or rake the surface to remove any weeds that have survived. Often we will find that the cardboard has completely broken down and the area is ready for use.

ROTATIONS

This is the term given to growing crops in a different place in the garden each year. It is one of the cornerstones of organic practice, helping to balance the needs of plants and the pressures of weeds, pests and diseases. The theory is that different crops require different nutrients, and are susceptible to different pests and disease. If we grow a crop in the same place every year, there is a risk that we allow a deficiency of a nutrient, or let a particular pest get too numerous. Some plants are also better at smothering weeds than others: for example, the large spreading leaves of a cabbage compete better with weeds than the thin upright onion. We can keep weeds under control to some extent by changing the crops we grow.

Not everyone agrees rotations are needed. Some argue that if you get your soil biology going well and save your own vegetable seeds you can successfully grow the same crops in the same place every year: for example, the Shumei Natural Agriculture approach. I see rotations as insurance; most of us don't have a perfect system and we can reduce our risk of failed crops with good rotation.

There is no perfect rotation but some basic principles can help us plan our plantings each year. Look at plant families first. For instance all the brassicas suffer from similar pest and disease problems, so you can group cabbages, sprouts, cauliflowers, rocket and turnip together in the rotation. Similarly squashes, courgettes and cucumbers go together. Once you understand which plants go with which, you can identify high-risk families and allow a few years between planting them in the same place each year.

Diseases that survive long term in the soil pose the highest risk. It is very hard to get rid of something like onion white rot, clubroot or sclerotinia in your soil. Onion white rot for instance can persist in the soil even where no host crop is grown for fifteen years. Crops that are in the ground for a long time are also more likely to suffer from recurring problems than those that mature quickly. For these reasons we should pay particular attention to the following crop groupings: Allium, Brassica, Potato.

Protected cropping rotations are a particular challenge, as so many of the crops we grow under cover belong to the same family. Tomato, pepper and aubergine for instance are all members of the Solanaceae family.

Four years is the normal recommended minimum for a rotation to give a good break between these high-risk crops; longer rotations are even better. Once you start planning your own rotation you quickly realize that some crops take up more space that others, and you'll want to grow lots of your favvourite vegetables. Juggling the space and health requirements of all your crops can be complicated. I try not to get too hung up on my rotation but just leave as long as I can between families within the practicalities of space and need. A focus on soil health and a healthy garden ecosystem reduces the risk to individual crops.

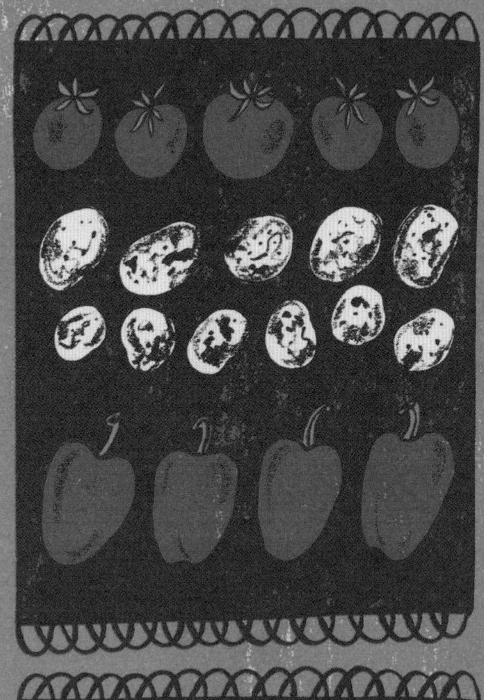

CAPTURING SUNLIGHT

All food-growing relies eventually on the sun – plants capture sunlight and use it to make sugars from water and carbon dioxide. Our aim as gardeners is to grab as much of sunlight as we can, to grow the maximum amount of food. Sunlight that hits bare soil rather than a growing plant is wasted.

Many of the inputs we use to help plants grow are based on sunlight. Man-made nitrogen fertilizers rely on oil and gas made from plants over the millennia. Animal manures also derive originally from sunlight, because animals eat plants, which need sunlight, and excrete the manure for us to use to feed our garden. The more of the current sunlight we can capture, the less we rely on using our bank of stored energy like gas and oil.

A core principle of organic gardening is to keep the soil covered for as much of the year as possible. Our ideal is to have a plant growing at all times to capture energy. However it is not always possible to have a food crop growing so we can also use other plants to fit into our garden to help. Non-food crops grown for this reason are often called 'green manures', and there are a number of ways we can use them in our rotation and planning

Sunlight that hits a plant is transformed into food and stored.

Sunlight that hits the soil is reflected as either wasted light or heat, so plant cover crops to reduce waste.

GREEN MANURES

The principle of a green manure, or cover crop as it is also called, is to maintain and build fertility and soil health. Bare soil is at best wasted light and water and at worst can lead to damaged soil structure and leached nutrients. Green manures help to hold on to existing soluble nutrients, and build carbon and nitrogen levels in the soil.

All kinds of plants can be grown as green manures, though they are often not the same ones grown to eat. Even weeds can be used as a green manure, but many cover crops have been bred for a particular purpose. Clovers fix nitrogen, grasses build carbon in their fibrous roots and chicories have deep roots that break up compacted soil and bring up nutrients from lower soil levels.

SHORT-TERM GREEN MANURES

Usually very fast-growing annual plants, you can use them for just a few weeks' gap between crops, though they can last longer. They protect the soil, collect energy and soak up soluble nutrients in the soil that might otherwise leach out through rain.

Phacelia, mustard and buckwheat are great quick-growing examples of short-term green manures. They fit when the gap between crops is not long enough for another vegetable but too long to leave the soil bare, after early-harvested crops like potatoes or summer carrots, say. You might also pop them in after purple sprouting broccoli or leeks, which finish in early spring, before planting out a crop like squash or beetroot in early summer. The green manure germinates and grows speedily but is soft and easy to cut and incorporate back in the soil before the following crop, or you to remove for composting.

LONG-TERM GREEN MANURES

These plants are for extended periods between crops, or for building fertility during a fallow period. These longer 'leys' are used less in home gardens than in professional growing, but can still be a great way to give your soil a rest and prepare it for heavy feeding crops like potatoes and brassicas, which tend to be in the soil for at least six months, sometimes up to a couple of years. Green manures left in for a long time need to be cut back (strimmed or mowed) to keep them productive.

I recommend a diverse variety of plants within a long-term green manure as evidence shows that the overall productivity of mixtures outperforms even the most productive single variety. Long-term mixes usually contain legumes such as clovers or tares. The relationship between legumes and a soil-living bacteria allows them to fix (or capture) nitrogen from the air and turn it into a usable food for plants, as they cannot do that for themselves. These legumes can also be mixed with grass species that produce big fibrous roots and are good for building soil carbon. You can throw in a few other species like chicory or tillage radish, which have very deep roots and are great for bringing up nutrients from lower in the soil and helping to reduce compaction and improve drainage.

INTERPLANTING *and* UNDERSOWING

Interplanting

Plants grow at different rates and are harvestable over a long period. We can take advantage of these characteristics with some clever planning. When we plant out a large, slow-growing plant like cauliflower we need to leave enough space to allow for its final size, but this means there is a large area of wasted ground between the plants for a month or so as it grows. Using super-speedy growers like radish, rocket or lettuce, means we get an extra crop from the same area, and helps to reduce the wasted light and water that otherwise would have fallen on that bare ground.

This technique is especially useful in smaller gardens where every centimetre counts, but can be applied to any system and is a way of reducing the amount of work. You can get more crops from less land, meaning saved labour and more efficient use of resources.

Companion planting is another method of interplanting. Here we take advantage of an aspect of one plant, such as its smell or its ability to attract predators, to help another plant. For instance members of the allium family can be interplanted with carrots to help disguise the smell of the carrot leaves and keep the carrot fly confused. Or we might plant a flowering plant among our squashes to attract more pollinators.

Undersowing

This is sowing a green manure crop beneath a food crop. When it works it is very effective, giving weed control, soil protection and soil health benefits. It requires a balance between the size of the crop and the undersown planting to ensure that one does not dominate the other. Results can vary from soil to soil and even year to year, depending on temperature and rainfall.

Crops to try this on are large brassicas like sprouts and kale, and courgettes and squash. Choose a short spreading plant; usual choices are yellow trefoil or white clover. In warm wet years you might have to mow the undersowing to prevent it swamping the main crop. If you have a light soil and are in a dryish climate you can try one of the more vigorous clovers as they are less likely to take over. There is an element of trial and error as you discover what works for your soil and cropping system.

I've had some success undersowing tomatoes and climbing beans in a polytunnel. You need to treat them as a crop and keep them watered; they will grow taller than they would outside.

Also, once the main crop is finished you have a cover in place to build fertility immediately without having to do any more work. Some crops like courgette can even be planted into an existing stand of trefoil, just dig out a hole about 60cm/2ft in diameter and plant in it. The plant should grow quicker than the green manure, and be able to compete.

YIELD

When starting out, it's pretty hard to know how much of each crop to plant. How do I choose which plants I can fit in my garden? How many courgette plants will I need? How much space do I require to grow my potatoes? Even with the internet, it is not easy to find yield and spacing information, which is one of the reasons I decided to write this book.

Don't be disheartened if you don't get the predicted yield from your crops as there are so many factors affecting how well they perform. I have been conservative in my yield estimates, and in many cases you may be able to exceed them, but if you don't we can always blame the weather.

SOIL

Your soil type has an impact on how your plants grow. Asparagus prefers a light sandy soil, while cabbage likes a heavier clay soil. You can still grow most crops in most soils but you may not get maximum yield from them. How you manage your soil makes a difference. Building soil health will give healthier crops that grow better and yield more.

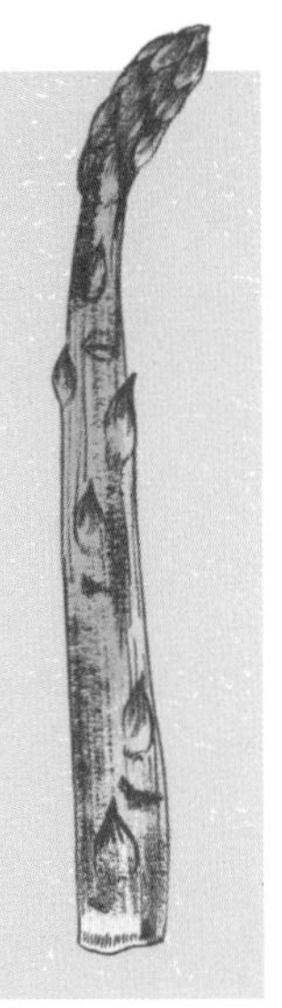

WEATHER

I remember early on being very despondent about an onion crop, thinking I had messed up and was probably no good at growing vegetables. At a growers' conference later in the summer everyone talked about how bad the year had been for onions. The weather in a given year will suit some crops better than others. Take the good with the bad, and accept some write-off crops each year.

As our weather becomes less predictable with climate change we also have to be willing to ignore some traditional advice on timings. Late frosts are rarer in the UK for instance, though still not unheard of, so planting out tender plants like squash in late spring might now be worth a gamble where fifteen years ago I wouldn't have risked it.

Growing in a polytunnel or greenhouse also greatly affects yield. For some crops, such as aubergine in cool climates, it will be the only way to harvest anything at all, while in others it may extend the time of year we can grow crops such as early beans or winter salads. Oriental brassicas like the cool weather but aren't so keen on the wind – the best yields I have had from these crops have been under cover in early spring.

VARIETY

There are significant differences in yield between varieties. Crops are bred for many characteristics: yield, flavour, pest or disease resistance, colour or size to name a few. I rarely choose a variety purely for its yield potential and am usually willing to sacrifice crop weight for taste.

ONE-PICK CROPS

Many crops have a single pick, when everything from one plant is ready for picking at once. Potatoes, onions and apples are examples of this. Sometimes something that happens earlier in the season will have a dramatic effect on the crop, such as water stress or an unnoticed pest attack.

SKILL

When we start anything we make mistakes, that is how we learn. I have lost many crops but have also had some astounding successes. We tend to credit our successes to soil and weather and blame our failures on ourselves. Keeping a record of your growing is a great way to help learn from both the good and the bad. If you know when you sowed and planted out a crop that did well, and how hot and wet it was that year, you have a better chance of repeating your success.

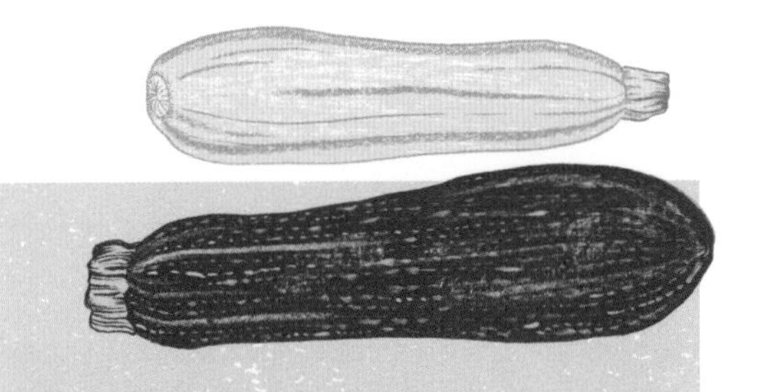

MULTI-PICK CROPS

Lots of fruit and vegetables produce a crop over a longer period of time: peas, courgettes and kale for instance. For many crops how and when we harvest can have an impact on total yield. If we remove the growing point of cut-and-come-again salad varieties by chopping too low we may stop them producing. Conversely we can harvest the lower leaves individually, which can help reduce slug and disease damage by lifting the productive part of the plant and allowing air movement through the crop. If we stop picking climbing beans they will enter the maturing, seed-producing phase and stop producing beans – in other words the more we pick the more we get.

I prefer to eat small tender vegetables, so I pick my peas and beans small. This means I get less overall yield. The size we harvest courgettes will also have a significant impact on overall crop weight. Do we want massive marrows or small courgettes? If we harvest for the flower on courgettes we get even less weight. We can choose.

HOW TO CHOOSE WHAT TO GROW

One of the likely limiting factors in what you might grow is how big your garden is. Some crops such as herbs and salads can be squeezed into the smallest of spaces, while potatoes, globe artichokes or fruit trees are likely to be an option only for those with a bit more space. Unless you are aiming to be 100 per cent self-sufficient and have unlimited space, you have to make some decisions about how much you can fit in. One of the main aims of this book is to give the information needed to be able to properly make those decisions; however it's also worth thinking about some basic principles.

HOW BIG DOES A PLANT GET?

I remember being surprised the first time I grew a courgette plant by how far it can spread in a short space of time, and yet the yield from it is relatively small. A lettuce plant on the other hand is tiny by comparison but we eat the whole thing, so arguably is a better use of space. The plants of which we eat the whole leaf are likely to give the best return for space, but we can't live on leaves alone.

THINK VERTICALLY AS WELL AS HORIZONTALLY

Climbing plants are a good way of increasing your cropping area. Squashes, beans, cucumbers, tomatoes and fruit (both soft and top fruit) can all be trained upwards to make use of light.

EXTREME EXAMPLES OF BIG SPACE SMALL REWARD PLANTS

✚ **Globe Artichokes** A beautiful easy-to-grow large plant, not totally frost hardy, this gives scant, though delicious, return for space. Only add to your list if you've more space than you know what to do with.

✚ **Asparagus** This perennial favourite grows quite large, doesn't do well with weeds and crops only for a short season.

✚ **Cauliflower** Hard to grow consistently well, often a small head is nearly ready one day and gone over a couple of days later. Grow purple-sprouting broccoli instead.

EXTREME EXAMPLES OF SMALL SPACE BIG REWARDS

✚ **Most herbs** You don't need a lot of herbs to go a long way. Even herbs grown in a pot on the kitchen windowsill can make a useful contribution to your food. Most are relatively easy for novices to grow. Finally, they are expensive to buy.

✚ **Salads** Particularly cut-and-come-again salads do well in pots or small plots, and offer supplies of fresh leaves potentially year-round (if we choose the right variety and have some indoor growing space).

MANIPULATING SPACING FOR SIZE OF CROP

Basic principles

Most plants have a maximum size they grow to. Given sufficient light, water and nutrients they will reach this size and crop to their potential. We have also bred different varieties for a range of purposes, often with size as a key feature. Hence tomatoes will range from a small cherry to a large beefsteak. However breeding is not the only thing that determines what size a vegetable will grow to, and for some crops you can manipulate the spacing at which you plant them to reduce the size. I have used this method quite a lot in my growing career, supplying speciality produce to high-class restaurants because chefs like small vegetables as they can make them look good on the plate.

In general this technique does not work with plants where we eat the fruit or seed. For those plants close planting would merely mean less yield per plant (though overall yield per square metre/yard might not be affected). It does work for most root crops and for those that form a head or flower such as cauliflower and cabbage. One thing to watch out for though is that the plants are a bit more stressed, competing with each other so will be more inclined to bolt, particularly in dry weather.

Here are a few examples of ones I have either tried myself or seen growing elsewhere. However there are others that could be tried.

CAULIFLOWER

This is a crop often left to specialist growers and I must confess was not my most successful when growing commercially. I often ended up with small heads even when I was aiming for large. Unless you have a large family to feed, a large cauliflower can be a daunting proposal. Growing twice as many smaller ones can suit very well. Standard planting distance for cauliflower is anything up to 75cm/30in depending on the variety. Try planting at around 20cm/8in spacing for mini ones. This spacing works well on cabbages too.

SUNFLOWER

This is one that I haven't tried but saw working perfectly on a visit to Embercombe in Devon, England. Plants that normally grow to 2m/7ft tall with big heads were instead only 1m/3ft tall, with flowers the size of a tennis ball, which are more flexible for the average flower vase. If growing for seed it would not affect seed size.

CARROT, BEETROOT AND TURNIP

For root crops sow the seed more thickly than the recommended rate (perhaps one seed per 1cm/½in), and start picking as soon as the root is big enough to eat. As you thin out the baby roots it will allow the others to grow a bit more (each seed will tend to grow at a slightly different rate unless you are using an F1 Hybrid seed). You can eventually thin the crop out to normal spacing and allow the remaining roots to formal standard-size vegetables, or you can pick them all as babies and plant something else afterwards.

Don't try this on radish – it just bolts too quickly and any stress from overcrowding will mean the roots don't develop at all.

Breeding is not the only thing that determines what size a vegetable will grow to, and for some crops you can manipulate the spacing at which you plant them to reduce the size

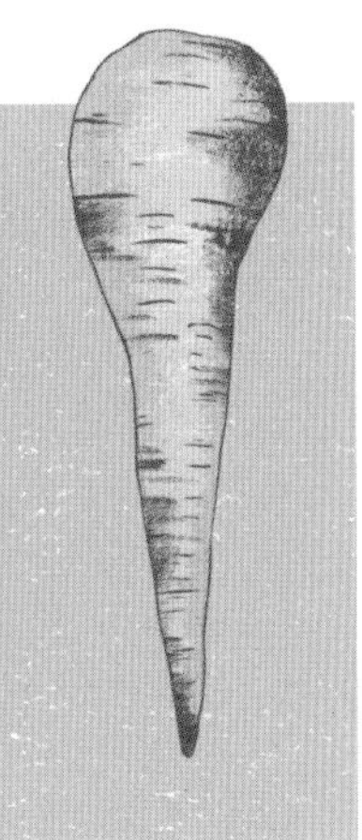

PARSNIPS

These need to be left for harvesting all at the same time so you can't sow quite so thickly but you can still pack them in more closely than recommended (for instance 5–10cm/2–4in spacing rather than the 15cm/6in normally given).

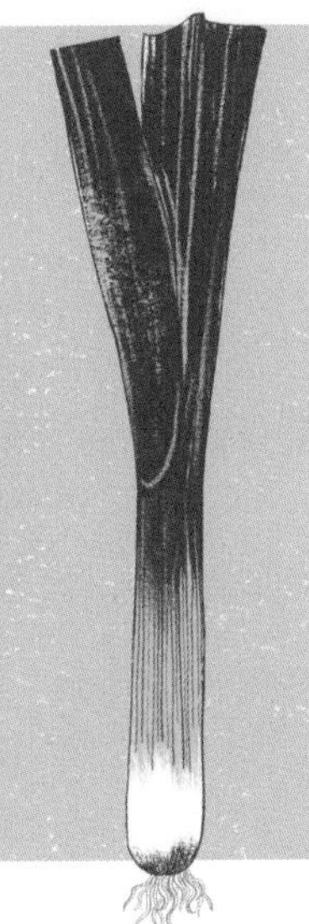

LEEKS

Another chefs' favourite. I used to sow leeks directly into rows (this is less usual, as for leeks transplanting is more common) aiming for each plant being 1cm/½in apart. I would then harvest these once they were about as thick as my finger. Steamed or braised whole they were delicious and looked lovely when served.

Chapter two

TASTE

Growing fruit and vegetables for me is rooted in a love of food. I also hate waste and want the most from my garden for the least effort. But growing is only part of the challenge. When to harvest and what to do with your crops if you are not going to eat them immediately are key.

We have become conditioned as to which bit of plant we use. Plant parts that are a bit fiddly or tricky to pack have become 'waste'. Lots of plants produce edible flowers and seeds that aren't commercially viable but can be enjoyed at home.

Knowing how to keep harvested crops in great condition for as long as possible reduces waste. Rediscovering traditional food preservation methods, but adopting new technology to make them even better, is vital for doing this, and is important, too, for gluts of produce harvested for relatively short periods in summer and autumn.

Drying is wonderful in hot dry climates, but it's hardly zero waste if you have to run your oven for days in cooler wetter conditions. You can make the most of dry summer days with a solar dryer, or tap into waste heat sources at home.

Most fruit and vegetables freeze well, and good freezers use relatively little energy. Pickling and fermentation are even better: apart from initial blanching and heating of brine. Most can be stored at room temperature in reusable glass.

Many root crops can over-winter in zero-energy storage using techniques that were standard before we had fridges and freezers.

DON'T THROW IT, STORE IT: BASE RECIPES

However carefully we plan our growing, every year we will have too little of some things and too many of others. That is the joy of working with unpredictable nature. The key is to be flexible in using what comes out of the garden.

Here are a few very simple-to-make, basic recipes that can be adapted to make use of whatever you have to hand. I have given a few examples of alternative ingredients but they are by no means exhaustive.

Tomato sauce

This is so good when made with fresh tomatoes that it can be eaten as a thick soup or be used as a base for a range of pasta dishes, added to stews or reduced a bit further for pizzas. If you don't have fresh tomatoes using tinned works well too.

YOU WILL NEED

- 2 tbsp olive oil
- **Onion bit** – 1 onion/1 leek/1 bunch spring onions/4 shallots
- **Root bit** – 1 carrot/1 small parsnip/1 beetroot/50g/2oz squash
- **Celery bit** – 1 stick celery/50g/2oz fennel bulb/50g/2oz celeriac
- **2 garlic cloves** – not really an alternative to this one, but you can always add more of the onion bit to compensate if you have no garlic
- **Herby bit** – 1 bay leaf and 1 teaspoon chopped sage/rosemary/oregano; you can add pretty much any herb to this sauce but I tend to go for stronger aromatic ones
- **750g/1.5lb tomatoes** – of course any tomato will do. I like to quickly roast or grill them first to give an extra richness by laying them out on a baking tray and cooking under a high heat for 15 minutes

METHOD

Just finely chop all the ingredients and fry everything except the garlic and tomatoes gently until they've softened (10–15 minutes on a medium heat should do it). Then add the garlic – it can burn if overcooked. Then put in the tomatoes and cook on a low to medium heat until the mixture has reduced to about half of its original amount.

Smoothies are a great way of using gluts of fruit. They work particularly well for soft fruit which doesn't keep so well.

Fruit smoothies

Smoothies are a great way of using gluts of fruit. They work particularly well for soft fruit which doesn't keep so well.

YOU WILL NEED

✚ **Fruity bit** – any soft fruit such as raspberries, strawberries or currants; to give a bit more substance add a banana, or a couple of spoons of oats.

✚ **Creamy bit** (optional) – half fruit half plain or Greek yogurt or milk works well too, if you like your smoothie a little creamier.

✚ **Sweet bit** (optional) – 1 teaspoonful honey or maple syrup.

METHOD

There is no magic art to making a smoothie. Just pop a mix of ingredients to suit your taste into a blender and mix them up. However, here are a few things I've found that help:

Make sure there is enough liquid. If too dry it won't mix properly or be drinkable. You can add milk, water, yogurt or any liquid you like. Some fruit and vegetables are wetter than others so take each batch as it comes.

You can use a few chunks of fruit or vegetable straight from the freezer, but too many frozen ingredients can hinder the blending process.

I don't have a sweet tooth but if you do, add honey or maple syrup when using tart fruit like currants.

Herb pesto

Though traditionally made with basil and pine nuts there are other leafy herbs and other nuts/seeds you can use to make pesto. Italian pesto includes Parmesan cheese too. I have left that out of this recipe but you can add it (about 50g/2oz) or an alternative like Old Winchester cheese. Pesto is great just to add to pasta for a quick dinner, but I also pop a few spoons of it into other dishes to pep them up.

YOU WILL NEED

✚ **Nutty bit** – 50g/2oz sunflower/pumpkin seed or walnut/hazelnut

✚ **Herby bit** – 100g/4oz basil/coriander/rocket/parsley

✚ **Oily bit** – 150ml/5fl oz olive oil or cold-pressed rapeseed oil. You can add small amount of walnut or hazelnut oil for extra nutty flavour

✚ **2 garlic cloves** – pesto without garlic is not quite as good but you can add shallots/onion/chives if no garlic is available

METHOD

Blitz all the ingredients apart from the nutty bit till you have a smooth paste. Separately grind the nuts or seeds so they are small but not ground to a flour. Mix the two together.

Vegetable stock

You can make big quantities and freeze in batches for use later. The beauty of making stock is that you can use the older, tougher bits of veg that you might normally not eat.

YOU WILL NEED

✚ **Veg bit** – pretty much anything goes here. Leave out the cabbage family but otherwise whatever is coming out of the garden can go in the pot. The traditional 'holy trinity' of stock is carrot, onion and celery. Onions needn't be peeled. Carrot tops, celeriac leaves and old veg that's gone soft but not yet rotten can all go in.

✚ **Herby bit** – rosemary/bay/thyme

METHOD

The method is easy too. Pile all the veg into a large pot, add some herbs, just cover with water and simmer over a very low heat for a couple of hours. Sieve and use in stews, risottos and soups.

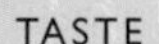

FREEZING

The key to freezing is capturing that fresh flavour. Freezing doesn't entirely stop decay, it just slows it right down. The quicker you can get produce from the field to the freezer the better. As with any storage methods, freeze the best produce, and eat the worst immediately.

Enzymes in our fruit and vegetables can affect flavour, colour and nutritional loss when we freeze food. There are a couple of steps you can take to deactivate the enzymes and reduce these losses in some foods.

For most green vegetables blanching before freezing will inactivate the enzymes, as well as kill off other microorganisms that might spoil them. You don't want the produce to cook though, so after blanching plunge the vegetables quickly into iced water to quickly cool. It's not essential to blanch before freezing but it will increase how long the food keeps. For green vegetables like kale and spinach, stir-frying can also work well and often gives a better flavour once reheated.

Try to exclude as much air during the freezing process as you can to help reduce the oxidizing of the fruit on the surface. In practice this means that a full small tub will keep better than a large, half-full tub (as well as being a better use of freezer space). Squeeze as much air from freezer bags as possible.

'The quicker the freeze the sweeter the peas.' As we freeze fruit and vegetables we are actually freezing the water inside its cells. A long slow freeze will give fewer larger ice crystals, and is more likely to damage the cell walls of the crop and result in reduced quality. A quick freeze gives lots of tiny ice crystals, which are less likely to break the cell walls. This is why we should cool our produce as much as possible before freezing, and also freeze in small batches. Domestic fridges are not designed to freeze food very quickly, and filling them with unfrozen (or worse, still warm) food overloads them and gives poor quality food.

I lay berries on a tray individually to freeze. This allows each fruit to freeze quickly, but also stops the berries sticking together, making it easy to use a few at a time. You can also cook up the fruit to make a compote and then freeze that in portions.

For food that we are going to use in soups and stews, I find that cooking before freezing is an effective method. You can batch-prepare tomato sauce for example (see page 32) then cool and freeze in portions.

A great way of freezing herbs is to put a few leaves into an ice-cube tray and cover with olive oil or melted butter. Once they have frozen, pop them out and put into a bag or container. They'll be perfect to add to pasta sauce or savoury pancakes.

Freezing in small-portion-size containers or bags makes food easier to use. Portion size of course depends on how many of you there are and how big your appetite is.

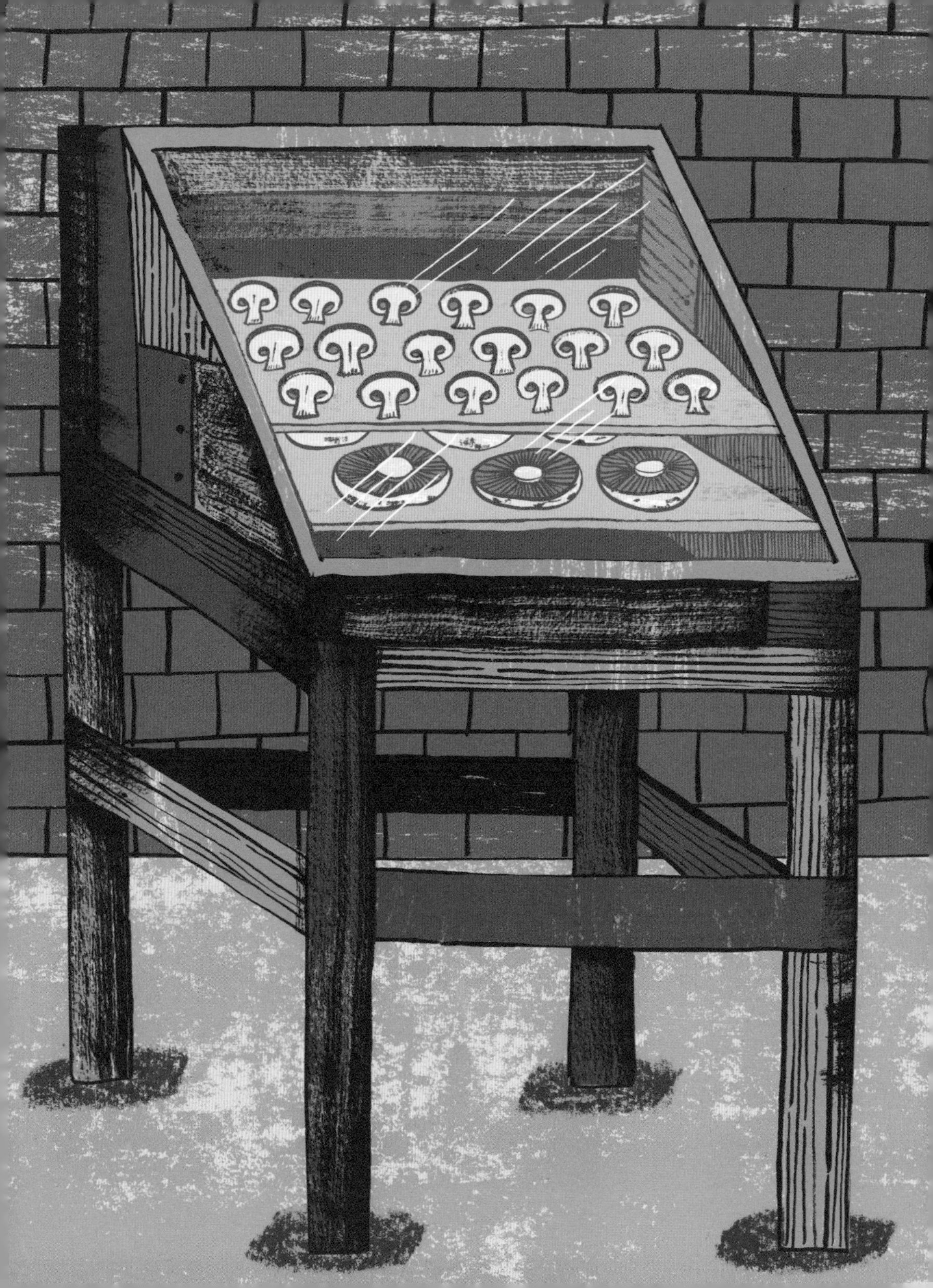

DRYING

Unlike freezing, which retains the water in produce, drying aims to remove the water. This slows the enzymes right down, meaning they will last for much longer. We also aim to stop mould and bacteria from destroying our crops. The drying process will change the flavour of most foods, though also intensify it.

The trick is a slow dry in low heat with lots of air. Dry any food too hot and the outside hardens before the inside is dry, trapping moisture inside. Food dried too quickly won't last. Most food will take at least six hours in an hydrator, and longer in an oven.

The best zero-waste way to dry food is using a home-made solar dryer from reclaimed materials. All you need is some wood and an old window or clear plastic sheet. Here's one example:

This method of course works best if there is lots of sun. Cool night temperatures are not great as they tend to bring moisture back into the system. Drying food like this will take a lot longer than using a hydrator or oven, up to four days, depending on heat and humidity.

If you don't have sun or the space for a solar dryer, a proprietary hydrator is your next best option, and they use a lot less energy than drying in an oven. However if you are drying only a small amount, embedded energy in the dehydrator might outweigh electricity savings.

If there is a suitable dry warm spot out of the sun, then air-drying is also an option for some crops like herbs and peppers. Just hang in bunches with a paper bag around them. Airing cupboards are a great zero-waste way of harnessing the 'waste' heat from a water boiler in order to preserve food. If there is space you can even hang the bunches over the back of a fridge or freezer, to make use of the hot air rising from the elements that keep the fridge cool.

Sun-drying mushrooms increases the levels of vitamin D significantly, storing it for you to top up your levels during the darker winter months

PREPARING YOUR FOOD FOR DRYING

For quick and efficient drying you need thin slices of fruit or vegetables, up to about 0.5cm/¼in thick. As with freezing, a quick blanch deactivates the enzymes and improves the quality. Spread the slices out well so none of them touch each other.

Let dried food cool right down before storing, to prevent condensation from spoiling it. However, leave it too long and it can absorb more moisture from the air. Usually 45 minutes is about right, depending on the size of the pieces.

Properly dried crops can last a long time. Keep in airtight containers: glass kilner jars are perfect, or plastic containers. Cool, dark storage areas help preserve both colour and flavour.

USING ALL THE PLANT

One of the easiest ways to reduce waste is to eat more of what you harvest. While I have no problem putting bits of plants that I don't eat into the compost bin, I would still rather eat them if I can. Many plants have edible bits that are not normally consumed, either because they are not as tasty as the parts we normally eat or because we have just forgotten that we can.

Here are a few bits of plants that you might not have realized are edible. Most discarded bits of vegetables can go into the stockpot except for brassicas leaves ,which can make the stock smell a bit 'old cabbagey'.

BRASSICAS

Almost all of most brassica plants are edible, so even though we might normally just eat the flower of the cauliflower and broccoli, or the swollen stem of kohl rabi, we can actually eat the leaves, flowers, some seed pods (if they haven't become too tough) and even sprout the seeds. One great way to get some extra crop out of many brassicas is to leave them in the ground after the main cropping and harvest the secondary shoots.

ROOT TOPS

While we mostly eat just the root of beetroot and carrot, the leaves of both plants are perfectly edible too. Beetroot leaves can be used in the same way as you'd use spinach or chard, while carrot leaves are great in soups and stews. Young carrot leaves are even nice when added in small quantities in a salad.

Some leaves that feel a bit rough when raw, like turnip or radish, can be added to stir-fries, or sauteed in a bit of butter.

SEEDS

Though we don't always take advantage of them, many of our best-loved vegetables have edible seeds. Some squash and pumpkin seeds can be roasted and eaten, while chilli seeds can be added to your herbal teas for a little kick.

If space in your garden is not too tight, you can let plants that you don't harvest go to seed for eating: for example, fennel, carrot, poppy and sunflower are all worth growing for seed to eat.

APPEELING

I mostly like to keep the peel on my vegetables, unless it is really old and tough. But even if you prefer yours peeled, you might not need to throw the peelings straight on the compost heap. Here are a few uses.

✚ Onion, carrot and parsnip skins can be added to stocks and soups for a bit of extra flavour.
✚ Potato skins can be used to grow new plants, especially at the tail end of the season. I usually take a thicker than normal layer of peel off when doing this to ensure that each potato peel has enough energy to sprout.

PICKLING *and* FERMENTING

What is a pickle?

Pickles use the acid in vinegar, and salt, to kill the organisms that break down food. The other ingredients add flavour. Different vegetables need slightly varied treatments, but a few basic rules will make sure your pickles are safe and tasty:

✚ Check vinegar labels for strength: 5 per cent vinegar is essential. Use pickling vinegar as most salad vinegars are diluted. White is standard, but some cider vinegars are strong enough to use too.

✚ Stick to a tested recipe. Produce-to-brine (the pickling liquid) proportions are the key to proper preservation. Mix vinegar and salt, and water and sugar if using, before pouring over the produce.

✚ Boiling the brine first is not essential, but can help soften some vegetables. Boiling it with the spices might also bring out their flavours a little.

✚ Produce as fresh as possible is best. Discard any rotten or mouldy bits and wash it well. Don't expect good pickles from bad vegetables.

✚ To cook the vegetables or not depends on your own taste, and the vegetable. Root vegetables benefit from pre-cooking; beans and courgettes can get mushy, so are better pickled raw.

✚ To protect pickles for longer out of the fridge, heat up each jar to be filled with pickles, then seal tightly before the contents cool to make it airtight.

✚ It takes time for the flavours to mature. You can aste after a week and improve with spices.

What is fermenting?

Fermentation also relies on acid to preserve food, the difference is that instead of adding vinegar to produce, the beneficial bacteria 'lactobacillus' does the work. This bacteria is found all over the place, including in our guts, and also on the surface of vegetables. If kept away from air, lactobacillus naturally turns the sugars in the produce into acid, which we can take advantage of. Because it is an active process, fermentation changes the flavour more than pickling does.

The most famous fermented foods are probably sauerkraut and kimchi, but you can ferment pretty much anything, and it's relatively easy.

Wash the produce with a salt brine to kill off the unwanted bacteria. Lactobacillus is somewhat salt-tolerant so survives to do its work for us. It is vital to keep the air away from the food. This is easy if we're fermenting something heavier than the brine. However, sliced cabbage for instance tends to float, so we need to find a way to keep it under the surface, like putting a weight on top.

The fermentation process releases CO2 as the organisms break down the sugars in the food, so the jar must be opened regularly to release the pressure. Temperature affects the speed of the process, but room temperature is generally fine..

DAUCUS CAROTA
SATIVUS
Beta Vulgaris.
RAPHANUS Sativus

WINTER STORAGE

Before quick global trade allowed us to ship or fly produce around the world year-round, our survival depended on being able to store crops through the winter when our climate prevented us from growing much fresh fruit and vegetables. Though not currently a matter of life and death, you need to know how best to store food in order to make best use of everything you grow and know how to keep your autumn harvest tasty and edible for as long as possible.

As my Dutch grower friend Fred once told me: 'Your cold store is not a hospital.' Produce doesn't get cured, so only the best-quality fruit and vegetables should go in.

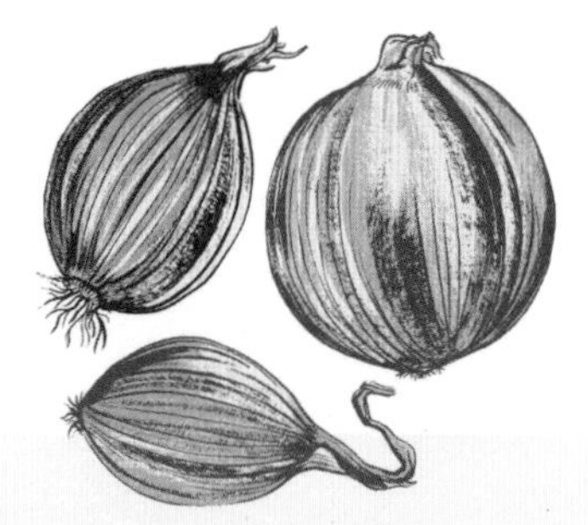

ONIONS

Onions need to keep dry and cool. The best way to store them is to hang them up in mesh bags in a cool shed. The even low temperature and air movement will prevent fungal diseases from taking hold.

FRUIT

As soon it is picked fruit is dead and starts to senesce (deteriorate with age). We cannot prevent this senescence; we can only slow it down. For instance with apples the main factors that affect this process are temperature, humidity and oxygen/CO2 levels. The organisms that ripen (or break down) the apples need to breathe and reproduce. If we slow them down by cooling the temperature and reducing oxygen levels we can keep the apples for longer. Most of us don't have specialist cold-storage facilities but we can still find a cool place to store them. Traditionally apples were individually wrapped in paper to help reduce oxygen immediately next to the fruit. Doing this also stopped disease spreading to the whole crop if one apple became rotten. I put my apples in a big sealed container in a cool shed and normally still have fruit in late winter.

There is a lot of variation between varieties – generally the later the variety the better it stores. Some varieties like 'D'Arcy Spice' or 'Tydeman's Late Orange' are even a bit better after they have been stored. Storage temperatures of 1–3°C/ 34–37°F are best for most varieties.

If we slow them down by cooling the temperature and reducing oxygen levels we can keep the apples for longer

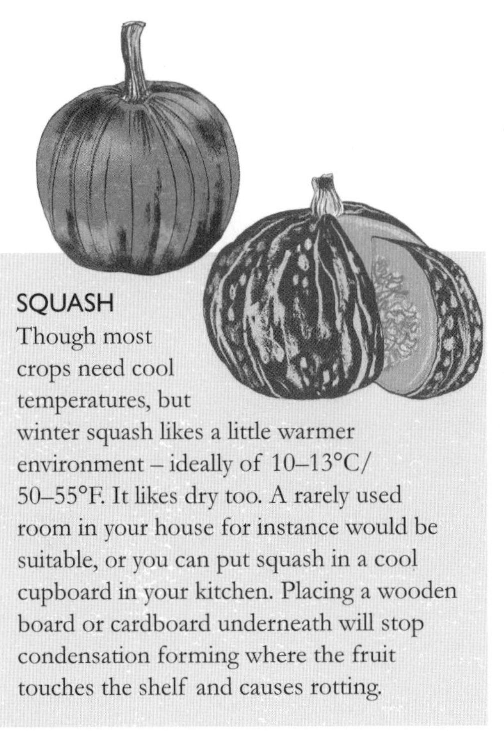

SQUASH

Though most crops need cool temperatures, but winter squash likes a little warmer environment – ideally of 10–13°C/50–55°F. It likes dry too. A rarely used room in your house for instance would be suitable, or you can put squash in a cool cupboard in your kitchen. Placing a wooden board or cardboard underneath will stop condensation forming where the fruit touches the shelf and causes rotting.

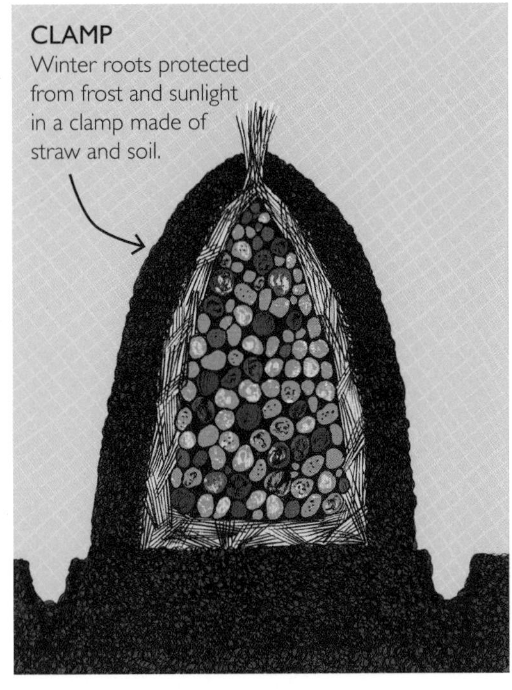

CLAMP

Winter roots protected from frost and sunlight in a clamp made of straw and soil.

ROOTS

Unlike fruit, most roots are still alive and wanting to grow. They are the storage organ for the plant over winter, before they reshoot the following spring. Many of our root crops are biennial. This means they last for two years, building up energy in the root during the first year, and using that energy to produce seed in the second year.

We want to keep the plant cold and dormant but still alive. We need to fool the root into thinking it is underground in a cold winter. While commercial growers use cold stores to mimic this, we can do it at home by creating what is called a 'clamp'. A clamp is a box or pile of material in which we bury the roots. Sand or compost is often used, though we can also just tightly pack the vegetables. The clamp is then covered with straw to provide insulation to keep the heap cool; sometimes a layer of soil is added on top of that.

You need to keep the clamp dry, so on heavy sites you may want to raise it up a bit.

Temperature and other environmental fluctuations speed up the deterioration of the crop. A north-facing place is ideal to prevent the winter sun heating up the store. If possible insulate your store: this might just be straw like inside the clamp or, if storing in a garden shed, even putting a couple of layers of bubble wrap around the produce to even out the day/night temperature difference.

Chapter three
WASTE

Our future on Earth relies on being able to live within our ecological means. While it can be disheartening to consider the enormity of this problem on a global scale, we can all contribute.

Organic systems aim to be holistic and self-sustaining. Most of us buy our compost, seed, pots, plants, plant food and tools. While this can be resource efficient, in many cases looking for alternatives can save energy and cost. Buy only what you need, and do only those activities that really achieve what you want.

In this chapter we'll look at how to get free fertility and how to reduce your use of plastic. Many gardening products are either made from, or come packaged in, plastic, so the more we can restrict these inputs the less plastic we'll have in our gardens.

Climate change means that for most of us rain comes less often, but is heavier. Trying to capture what we get and allowing our plants to cope when there is none are essential pieces of the Zero Waste jigsaw. Rainwater harvesting is one part, but building soil health and knowing how to water effectively and efficiently are also important. Choosing the right plant for dry or wet conditions helps too.

We will also explore how you can be more efficient in your garden, including no-dig gardens, sharp tools and careful use of old plant bits.

RAISING PLANTS

Most vegetables are grown from seed, while fruit is usually propagated vegetatively, getting them to form roots from bits of plant. New gardeners will likely buy seeds and plants, but as confidence grows it's fun to save seed, take cuttings and even try grafting fruit trees. The techniques are not complicated and it is satisfying to play a part in the whole cycle of a plant's life.

Raising plants from seed

You can sow seeds either directly into the soil, where you want the plant to grow, or you can sow them in a pot or tray and transplant the seedlings once they have germinated. Some crops, for example carrot and radish, don't like being transplanted, but for most vegetables there are advantages to transplanting. You can give the seedlings extra care until they are big enough to fight against pests (and especially slugs). Not all seeds will germinate. When sown direct you need to sow more than required and 'thin out' to the final spacing. Transplanting means you only plant out those that have thrived, thus saving seed.

CELL TRAYS

Also known as modules, these are a great way to raise plants for transplanting. You plant one seed per cell (or two and discard the weaker seedling); cell sizes differ to suit different crops. They reduce compost use, and robust designs last for years. The cells are small, drying out quickly in hot weather.

BLOCKS

Soil blocks work on the same principle as cells, but cell-like blocks of compost are formed with a block former. Many commercial growers use these, and young plants seem to suffer less from 'transplant shock' than they do from cell trays.

BARE-ROOT TRANSPLANTS

A zero-waste method of plant raising is to sow the seeds thickly in the ground. They will need protection early on; use a cloche if you don't have a polytunnel or greenhouse. Once the seedlings have at least two sets of true leaves, carefully pull them up and replant in their final position. Most brassicas and leeks work well this way, and in my experience outperform cell-grown plants. Plants raised like this can be grown larger before transplanting, which makes them better able to survive pest damage after planting out.

TEMPERATURE AND MAKING A GERMINATION BOX

Different plants have different germination needs – the main one being temperature. Propagators provide a suitable environment, or you can leave seed trays on an indoor windowsill to germinate. A germination box is most efficient for most vegetables: an insulated box with some form of heating in it and into which you put your seed trays. It will provide a constant heat and damp environment for minimum energy.

Some vegetables also need light to germinate.

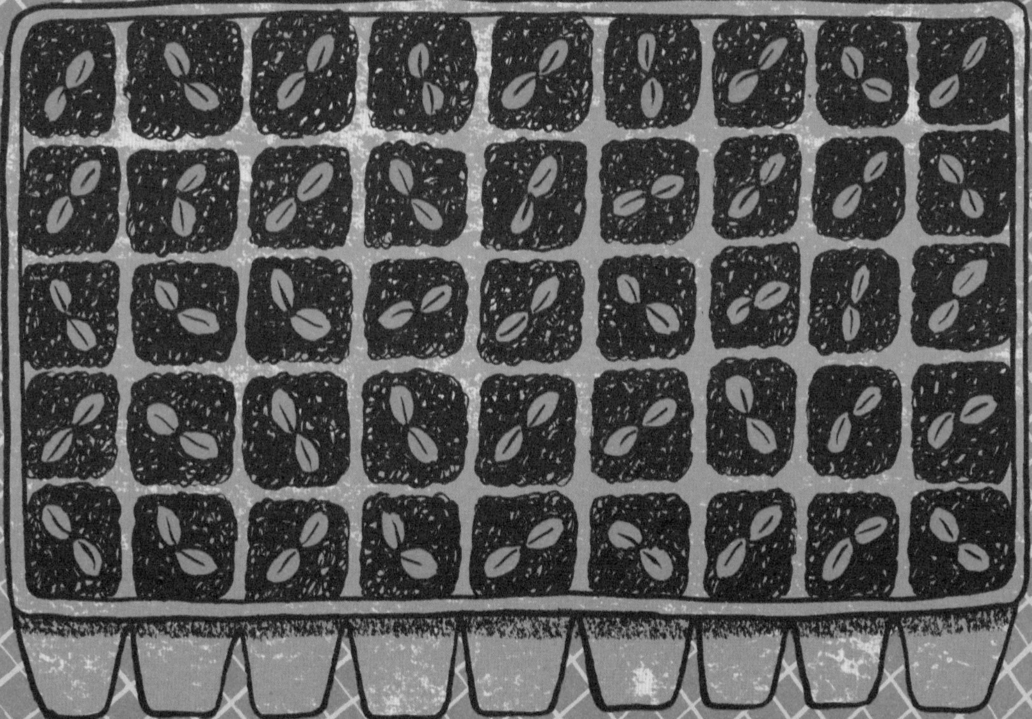

Raising plants from bits of plants

SAVING YOUR OWN SEED

This is a great way to reduce waste, unnecessary seed packaging and transport. As your skill builds you can even get healthier, more productive seeds that are adapted to your own growing conditions. Start with easy-to-save vegetables like tomato, peas or beans. They pollinate themselves so will come out as the same variety as the plant you grew. Just save seed from the healthiest, most productive plants.

DIVIDING

Perennial plants like rhubarb or Jerusalem artichokes need to be divided every few years to stay productive. In practice, this means splitting the root ball into smaller pieces, and replanting. For big roots like rhubarb you can do this with a spade; they are pretty tough so don't worry about being brutal as you hack the roots to pieces. For plants with a more fragmented root such as Jerusalem artichokes pick some nice bits of root and replant in a new place.

CUTTINGS

Many plants will produce roots from sections of shoot or stem. Blackcurrants and gooseberries work well from hardwood cuttings, that is, from a section of stem cut off in the winter and stuck in the ground. Woody herbs like rosemary and thyme grow from semi-ripe cuttings, which use the softer end-growth of the shoot. Check a propagation book for details of the best way to take cuttings for any particular plant.

RUNNERS

Strawberries, raspberries and blackberries send out new shoots that develop their own new roots. You can take these bits, plant them and thus create new plants from them.

- Strawberries grow 'runners', which you can root in the ground or in a pot.
- Raspberries have suckers, which can be dug up while dormant and replanted.
- Blackberries root from the tips of their vigorous shoots; once rooted you can cut each from the mother and replant it.

GRAFTING

Though this is a little more complicated, it is a lot of fun. Take a cutting from the fruit, for instance apple, that you want to grow. Then join it to a rootstock, usually chosen for its size or disease resistance. Splice them together. Eventually they will fuse to make one plant.

ZERO-WASTE FERTILITY

Old-school gardening books (and even some modern advice) will tell you to add a proprietary granular fertilizer to soil every time you plant. In most cases this is not only unnecessary but can be wasteful and even harmful to your soil.

Fertility comes from three places: the sun, the air and the soil. Plants use sunlight to take the nitrogen and carbon from the air to make carbohydrates. All other nutrients come from the bedrock from which soil forms.

Most soils have almost all the minerals needed to feed your plants. The biology of your soil just needs to be right to make them available to plants. The key to this is soil organic matter: building carbon in the soil. This is both very complicated and rather simple. Scientifically there are infinite interactions between plants, fungi and animals within the soil that if properly supported will do this. Practically, you can build soil carbon by adhering to three basic principles:

1 Disturb soil as little as possible – we have to cultivate it a little or we could not grow crops. However your guiding principle should be to do only the minimum needed to establish your crop.

2 Add organic matter – compost, manure, woodchip are all things you can use as a soil builder. Each has different properties but broadly they do a similar function from a fertility point of view, which is to feed the organisms in the soil and maintain or increase soil organic matter.

3 Never leave soil bare – at best, bare soil is a wasted opportunity to build fertility; at worst, it can lead to lost nutrients and carbon as bare soil is vulnerable soil. Avoid leaving soil exposed with undersowing, interplanting and cover crops.

In a zero-waste garden everything is a resource. All plant waste can be composted and returned to the soil as would happen in a natural system. If you cannot produce enough material yourself, find a good source of organic matter from elsewhere to supplement your own compost.

Woodchip has perhaps the most potential to enhance soil fertility. Many tree surgeons pay to get rid of their 'waste' chip and happily give it away if you have somewhere they can dump it. Don't dig woodchip into soil but lay it on the top as a mulch or, if you have the space, pile it up and compost it for a few months. You can make your own propagation compost from it by composting for longer and sieving. Trials show it performs as well as leading peat-based composts.

If your soil needs a helping hand, liquid fertilizers can be useful to aid a really hungry crop like tomatoes, or when you are still building up the health of poor or neglected soil. You can buy some great sustainable liquid feeds, but to go zero waste, you can make your own. Classic ingredients for liquid feeds are nettles (high in nitrogen needed for leaf growth in spring) and comfrey (lots of potassium, vital for a range of plant cell functions). Steep them in water to release the nutrients, and feed to hungry plants.

REDUCING ENERGY USE

Reduce machine use

Most of us rely on labour-saving gardening tools that use fossil fuels. Mowers and strimmers make jobs quicker and easier, though true zero-waste gardeners might aim to eliminate anything but hand labour. I'm not there yet, but here are ways to reduce your garden energy consumption.

Mow less often and leave grass long, cutting it only occasionally. This reduces fuel consumption, is great for wildlife and helps build soil health. Taller grass means bigger roots, which results in more carbon being stored. Hand mow small areas short, or get skilled with a scythe.

A lot of embedded energy is used to make a machine that may be unused most of the time. Sharing or hiring tools reduces your portion of that embedded energy.

Change your perspective to see chores as exercise. Chop young branches small enough to compost; or bash to break them to start the rot.

In almost all gardening situations a rotavator is unnecessary and damaging to soil – don't use one. To prepare for planting, it's much better to mulch with cardboard and woodchip and wait for nature to prepare the ground for you.

A plant will grow in almost anything that will hold compost, from old loo rolls to yogurt pots, and you can reuse plant producers' plastic pots too. Making, bagging and transporting compost uses a lot of energy, and you can easily make your own if you have a bit of space.

Minimize human energy

As commercial growers know, small efficiencies can have massive impacts when multiplied up. Similarly, for a home gardener, time and effort saved on one job can go into another. Below are some suggestions to help you.

Move the compost heap not the compost. Sites are usually heaped in a corner of the garden, so the compost must then be moved to the growing vegetables. Placing it where you plan to grow heavy feeders the next year means you can just spread the compost around. And escaped goodness ends up in your beds, not wasted.

Instead of composting twigs or finished flowering stalks, lay them on top of rows of freshly sown peas, beans or squashes to protect them against pests like birds and cats. The plants will manage to climb through such barriers. Lay big leaves from plants like rhubarb or comfrey on the ground around a growing plant to control weeds.

To get free plants for no effort, allow plants to seed. Some crops like chard, brassicas, winter purslane and lamb's lettuce self-seed all over if you let them; treat as quick baby salad crops.

Finally: keep your tools in good condition and, counterintuitively, have an extra set of some to save time going back and forth; spend time and effort designing your garden to suit your needs and site – it will help in the long term; and go easy on digging, which can be detrimental to soil health and releases carbon into the atmosphere.

THE ZERO-WASTE GUIDE TO TOOLS

A zero-waste approach can begin at the planning stage. You need less than you might think to garden. There are thousands of wonderful gadgets out there to spend your money on, but all of them take energy and resources to make, and may end up sitting on a shelf in your shed, or breaking after a couple of times of use.

WHAT DO I REALLY NEED?

This does depend on the size of your garden of course. For small gardens using no-dig systems you might get away with just a fork, rake, trowel, pocket knife and secateurs. For larger gardens you will probably need a wheelbarrow, a hoe (to speed up weeding), and perhaps a spade to plant larger plants and a shovel to help move compost around.

Most tools beyond this are a luxury or personal preference. For instance there are loads of different hoe designs. Some are specifically made for a particular crop like the half moon onion hoe, which can get in under the curve of the bulb. Others are of different widths to fit narrow or wide rows, or they have developed in a geographic area like one of my own favourites – the Manx hoe. You can sometimes save time and effort by having a range of hoes at your disposal, but equally you could probably manage with one.

SOURCES OF CHEAP TOOLS

I was horrified recently to see someone throwing a full set of hardly used garden tools away at the recycling centre. I managed to divert them into the back of my car but this is a sign of how cheap things are and how little we value them. Auctions (online and real) and car boot or yard sales are a great source of cheap tools. For basic hand tools the older the better, as they tended to be better made and last longer that most modern tools.

BUY GOOD QUALITY NOT CHEAP

I know this is easier said than done when on a tight budget, but cheap equipment is a false economy. It will almost certainly bend or break quickly. Secateurs are a classic case of this. I have had the same pair of secateurs for twenty years, so although they were pricey (£40) to buy they have had an annual cost of only a couple of pounds. A cheap pair (£10) will last only a year or two and will blunt more easily.

KEEP YOUR TOOLS SHARP

Blunt tools waste energy and put strain on the tool. When did you last put a new edge on your secateurs? Other tools work better if kept sharp, too. Hoes are much more effective and less work if you regularly sharpen them, and even spades benefit from being honed. Sharpening stones are easy to pick up at car boot sales, and old cutlery steels are great for sharpening secateur blades.

NEW HANDLE NOT NEW TOOL

Before throwing a tool with a broken handle away, check if it can be replaced. On older tools with wooden handles they usually can.

WHEN TO SOW/PLANT/ HARVEST/STORE

Once you get the hang of gardening basics, the next challenge is being able grow a wide range of crops, of the size you need and making them last as long as possible.

SPACING

This is not an exact science, but is nonetheless important. Most of us have limited space and want to cram as much as we can into it. However if you plant your crops too close they won't grow properly and you might get diseases as there's not enough air moving around the crop. Or plants might get stressed as they compete for light and water, which can result in them bolting (going to seed early). Too much space around them and you risk weeds swamping your vegetables.

You can play around with spacing though: for instance you can plant cauliflowers closer for a smaller head. Crops like radish and beetroot can be sown thickly and then thinned as you do the first harvest; plants that hadn't quite developed will then fill out once they have more space.

ONE-PICK CROPS

These are plants that tend to come all at once: many of the roots, squashes and potatoes for example. To get a long eating season, be clever with choosing varieties that mature at different times, and sow successions to be ready over months. Treat very quick-maturing crops like lettuce similarly. Also, think about storage for crops you can keep such as onions

MULTI-PICK CROPS

Beans and peas, as well as tomatoes and cucumbers, are plants that will keep giving if you can keep them healthy and productive. However, the longer a plant is alive, the more likely it is to get a disease or be eaten by pests. For some fruiting crops you also need to keep picking or the plant will go into its seed-producing phase and stop flowering. Most will need a lot of water and some extra feed (particularly if you are growing in pots) to keep them productive.

WHEN TO SOW AND PLANT

Sowing and planting is dependent on soil and air temperature, and some plants are 'photoperiodic' too, which means they need a certain time with either long or short days. Planting them at the wrong time of year, even if the temperature is suitable, will not work.

Temperature for seed germination and early growth is crucial. Plants classed as hardy (they can survive below freezing) will not germinate well at low temperatures. Brassicas are perhaps the hardiest, able to germinate at as low as 4°C/39°F. Other plants like peppers and tomatoes need high temperatures (18–25°C/ 65–77°F) to germinate well, and continuing warm temperatures to grow. .

ZERO-WASTE WATERING

COLLECTING RAINWATER

Rainwater is free, and better for plants than treated water. Gathering as much as you can is ideal. When it rains hard you can help reduce flooding elsewhere by collecting rainwater and stopping it flowing down into drains and rivers.

Collecting rainwater from buildings with existing drainpipes is easy: just divert the water into water butts and tubs. Use any container – old bins, baths or even dig a hole in the ground – to collect it. If you can keep the water covered to keep it clean, and stop the filters getting clogged if you use a pumped watering system.

IBCs (Intermediate Bulk Containers) are great for storing water as they are square and solid, so unlikely to fall over, and have a tap at the bottom making it easier to get the water out.

KEEPING WATER IN THE SOIL

The best way to save water is to keep as much as possible in the soil. Building soil organic matter will increase the capacity of your soil to hold water, meaning it will take longer to dry out. In poor and sandy soils try adding biochar to your soil to help it to retain moisture.

Mulches are another great way to hold on to soil moisture for as long as possible. Anything that shades the soil and reduces evaporation will work. Even though plastic mulches are very good at keeping moisture in, I am trying to move away from them in favour of something organic like compost, woodchip or even wool.

BEST FOR PLANTS

A good rule for watering soil-grown plants is to do so less frequently but with more water. If you water a little every day you will likely wet only the top few centimetres of soil, thereby encouraging the roots to stay in that area. Then if the surface does dry up, all the roots will die. If you make sure the lower soil levels are wetted, the roots will grow deep to find the water, and if the surface dries, plants will be fine as they have deep roots.

WATERING SYSTEMS

I try to avoid watering soil-grown plants, apart from when they are first planted and need watering in. This can work for many plants, if you have good soil with sufficient organic matter, and don't live in a very hot, dry climate. There are exceptions: delicate leafy plants like lettuce always need regular watering; and plants that produce regular fruit like beans or courgettes, need a good water supply through their fruiting period.

In most situations you could cope with watering cans. Hosepipes are quicker but water pressure is essential.

Drip and trickle irrigation systems are a low-labour and water-efficient way to water plants.

Chapter four

THE PLANTS

When you are starting out in gardening, whether with a few pots on a balcony, taking over an allotment or on a larger plot, working out what and how many of each plant to grow is tricky. Although the internet has made it easier to find yields for many crops, most sites and books will just provide standard spacings with no indication of how much produce you might expect to get from your plants.

This section aims to give you that guidance. Though not an exhaustive list of crops, I have selected a good range and given an indication of how much space they will take up in your garden to allow you to calculate how many to grow. As we have seen on page 20, these figures shouldn't be seen as gospel, there are just too many factors that can affect yield, but for planning purposes they are a great starting point.

I've also included some guidance for growing each of the crops. This is not designed to be a complete how-to for a particular crop – each fruit and vegetable could warrant a whole book on its own – but more to share with you tips from my own experience of growing and visiting the best growers in the UK over twenty-five years. This might be how much water each crop needs, or where there is a pest or disease to look out for, or perhaps which time of year each does best in.

For each crop I have also included how you can get the most reward from your efforts by maximizing the plant when growing and when harvested. You will discover which bits of plants you could be eating but probably aren't and how to continue to get the maximum value from a plant in the garden after you have harvested from it. Gardens are a complex ecosystem, and we rarely optimize the potential both for ourselves and for the wider environment. Leaving brassica plants to flower overwinter, or letting crops seed on bare ground as a cover crop, are just a couple of examples, but there's lots more. If you are a very tidy person, you may have to overcome the desire to keep everything immaculate. Zero-waste gardens can look wild but are still beautiful.

Finally, there are some tips on what to do when faced with gluts of produce, which will almost inevitably happen once you start growing your own. I take no responsibility, however, if you find yourself needing a new freezer, or your kitchen is overflowing with pickling jars full of your bountiful produce, or if your partner starts complaining about spare bedrooms full of drying seed heads.

ROCKET

Eruca sativa, E. versicaria

Salad rocket and its cousin wild rocket are super-easy to grow and a great fresh addition to salads and pizzas. If you like a bit of kick to your leaves then go for wild rocket. You can allow it to flower – the blooms are edible, beautiful and the insects love them.

Sow:	Early spring or late summer, direct into final growing space; early spring or autumn, under cover
Plant:	n/a
Harvest:	4–6 weeks after sowing
Eat:	Fresh in salad, or make into pesto for freezing

Yield

	Per Plant	Per m²/yd²
Total	100g/3½oz	1kg/2¼lb
Per pick	50g/2oz	500g/1lb

How often to pick

Pick every few days if harvesting individual leaves. If cutting the whole plant allow 2-4 weeks to regrow.

GROWING TIPS

✚ Rocket is really one of the easiest plants to grow. It germinates very quickly.

✚ Sow seed 1–5cm/½–2in apart in rows. After germination, thin the plants out, eating the thinnings of course and leaving one plant every 10cm/4in. Alternatively you can sprinkle seed over a patch, aiming for complete coverage. This helps reduce weeding.

✚ Harvest by cutting with scissors or a sharp knife. Don't go too low or you will cut out the growing point – aim for 3–5cm/1¼–2in above the ground.

✚ You should be able to get at least two cuts from each planting; in cooler weather you might manage three or four before the plant flowers.

✚ Like most brassicas rocket will germinate at low temperatures, making it a good early- and late-season salad crop. It doesn't like very hot or dry weather and tends to flower quickly in those conditions. Keep it well watered in the heat.

✚ Rocket is a great plant to intersow between slower crops such as sweetcorn or cabbages. By the time you have taken a couple of cuts the other crop will have grown up and started to shade it out. At that stage you can leave it in to attract insects, or if it is competing with your main crop remove it – picking off the last of the leaves and adding the stem to the compost heap.

✚ Unlike salad rocket which is an annual, wild rocket is a short-lived perennial plant, so don't dig it up once it flowers.

ZERO-WASTE TIPS

✚ The flowers are edible, and pretty.

✚ Even if you miss picking those you can also eat the young seed pods.

✚ Missed those too? Don't worry, allow the seeds to mature for a lovely free supply of rocket seed for your next sowing.

TOO MUCH ROCKET?

You will almost certainly have too much rocket at some point, especially once the leaves get big.

✚ You can add the bigger leaves to stir-fries or even wilt it as a spicy spinach alternative.

✚ The easiest way to deal with really big gluts is to whizz it up in a food processor with some oil and freeze it to use later. You can make pesto with walnuts and garlic, or just add to sauces and casseroles for a bit of extra bite.

ASPARAGUS

Asparagus officinalis

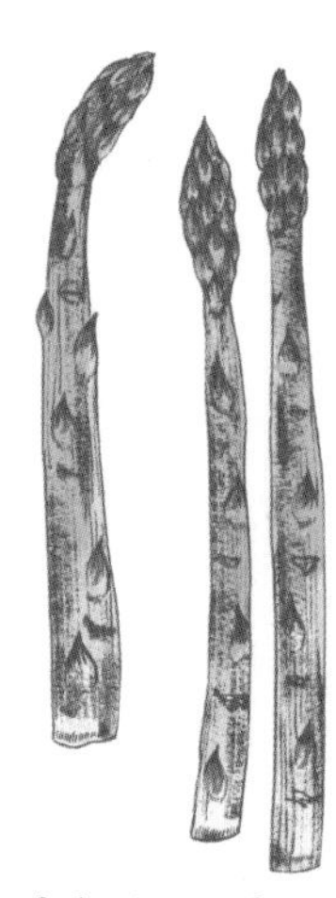

Truly this is one of the joys of spring. I could quite happily eat asparagus every day for its short, six-week season. In the right soil it will grow like a weed. If you are short of space it may seem like a luxury, but it comes when there is little else to harvest.

Sow: Spring, ideally at 21–29°C/70–84°F

Plant: Late autumn to early spring, while dormant

Harvest: Cut shoots just below the soil when they are about 15cm/6in long

Eat: Fresh, late spring to early summer; frozen/pickled, year-round

Yield

	Per Plant	Per m^2/yd^2
Total	500g/1lb	3kg/6½lb
Per pick	50g/2oz	300g/10oz

How often to pick

Pick every two to three days to ensure stems don't get woody.

GROWING TIPS

✚ Some varieties of asparagus are available as seed, and sowing your own is the cheapest option. However the more recent varieties (some of which are F1 varieties) are available only as 'crowns' (plants).

✚ Many new varieties are also all-male, which means they are more productive.

✚ Plant the crowns 45cm/18in apart.

✚ Using well-composted woodchips or straw, mulch the crowns overwinter to help stop annual weeds becoming a problem. If you start getting problems with perennial weeds then you can even put a thick plastic mulch over the crown from late autumn till the first shoot appears. It won't get rid of them but will slow them down a little.

✚ Asparagus likes light sandy soil and will tolerate salty conditions. If you have heavier soil then it is worth making a raised bed. This doesn't have to be a permanent structure but can be a ridged row, just enough to allow drainage and keep the asparagus roots dry.

✚ Asparagus doesn't attract many pests but does suffer from the asparagus beetle – a beautiful, black and orange creature. If you start noticing this then cutting the dead ferns back in late autumn will help to reduce numbers.

✚ Pick every two or three days for about six weeks in late spring and early summer.

✚ Don't pick in the first couple of years after planting; let the plants build up their strength.

ZERO-WASTE TIPS

✚ Though you can really eat only the young shoots of asparagus (once the buds open the shoots quickly get too woody to be edible), the ferns make a lovely addition to flower arrangements. Just be careful not to pick too many shoots, especially on young plants, as they need to grow their food reserves for the following year.

TOO MUCH ASPARAGUS

✚ I can almost guarantee you won't have too many shoots to eat fresh, unless you have a massive garden. However, in the unlikely event that there are more than you can cope with, they freeze well. Just blanch them, let them dry and lay them on a tray in the freezer to freeze separately. That way you can use as many spears as you like whenever you need them.

✚ Asparagus is also great pickled. Though in Europe it is normally the thick-speared, white variety that is pickled, green spears work well too.

CELERY

Apium graveolens

Growing blanched celery like you buy from the shops is a bit tricky, but makes a great gardening challenge. Self-blanching varieties are easier to grow though not quite so sweet. They tend to have a more intense flavour and are better for stocks and stews.

Sow:	Early spring
Plant:	Late spring or early summer
Harvest:	Before the first hard frost (100–130 days after planting), the head should be 5–8cm/2–3¼in at the base
Eat:	Fresh, or in stocks, soups and stews

Yield

	Per Plant	Per m²/yd²
Total	1 head (c.500g/1lb)	15 heads (c.7kg/15lb)

GROWING TIPS

✚ Celery is not the easiest plant to grow successfully. If you are new to propagating your own plants then consider buying in plants for this crop.

✚ Celery is a plant diva and likes to be pampered. It likes lots of food and water. Add compost before planting and make sure the plants don't dry out. If the plants are water- or cold-stressed during their early growth they can either produce too much leaf and not much stalk, or they often bolt and flower early.

✚ Make sure the transplant doesn't become root-bound as this can increase the likelihood of transplanting stress and early bolting.

✚ Try to keep young plants in temperatures of at least 13°C/55°F and don't plant them out too early in cooler climates.

✚ Celery can have quite a bitter taste, and one technique that growers use to stop this is 'blanching', which involves excluding the light from the stems to stop them producing chlorophyll. The stems are then paler and sweeter. There are various blanching techniques – for instance, piling soil around the stem – but the easiest is to wrap a paper or cardboard sleeve around easch stem, tying it carefully so the wind doesn't blow it off. You'll need to do this a couple of weeks before harvest.

ZERO-WASTE TIPS

✚ Celery leaves are great in small quantities in salads, or can be used as a parsley alternative in most recipes. They are also a fantastic addition to stock.

✚ If any plants go to flower before you can eat them (or end up too woody and tough to enjoy) allow them to go to flower. The blooms are a great attractant for insects, especially soldier beetles. You will then be able to harvest the subsequent seed. Use celery seed whole in a range of dishes; it is especially good with fish, chicken or in stir-fries.

✚ When you harvest celery, retain 3–4cm/1¼–1½in of stem in the ground. The plant will then reshoot and grow some more stems. These will be shorter and likely have more leaves than stem but are still delicious.

✚ You can even regrow a plant from the base of bought celery. Cut 5cm/2in off the base and pop into water until roots form, then you can plant it out into a pot or the soil.

TOO MUCH CELERY?

✚ Celery freezes fine – just chop into small pieces, blanch and freeze for use in sauces and stews.

✚ Celery also dries really well, intensifying the flavour. You can dry the leaves and stem (sliced thinly). If properly dried they will last for a long time and you can add them to stews and stocks.

CORN SALAD/LAMB'S LETTUCE

Valerianella locusta

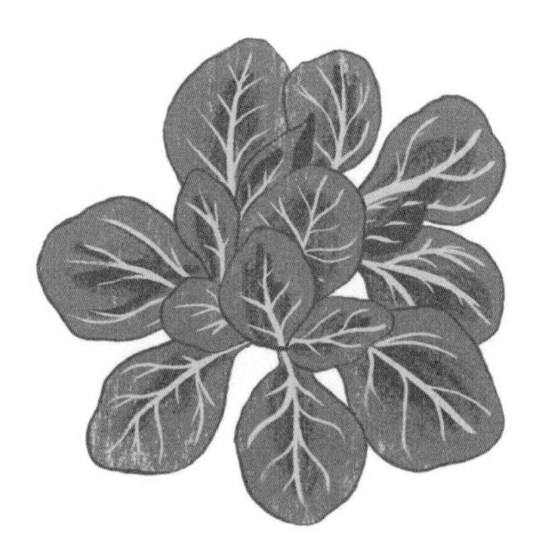

Lamb's lettuce is a fantastic little salad leaf with three times as much vitamin C as standard lettuce. It is frost hardy and you can grow it most of the year apart from the summer, when it tends to go to flower and seed.

Sow:	Late summer or early spring, direct into soil
Plant:	n/a
Harvest:	As soon as there are leaves to pick
Eat:	In salads

Yield

	Per Plant	Per m^2/yd^2
Total	25g/1oz	1.75kg/4lb
Per pick	n/a	n/a

How often to pick

Pick every week or two.
This interval will be longer in cold conditions.

GROWING TIPS

✚ Sow direct either in rows or as a dense patch. You are looking for a final spacing of about 10cm/4in, but you can sow more thickly and thin out to that spacing – eating the thinnings of course.

✚ Being hardy, lamb's lettuce will survive throughout the winter, though in temperatures below 5°C/41°F it won't actually grow much. You can grow lamb's lettuce in tunnels, though as with many winter salad crops keeping the tunnels well-ventilated even in cooler weather is essential to keep the crop disease-free. There is a tendency to think you need to keep all the heat in a tunnel, and though this is true at night and in windy or icy conditions it's good to throw the doors open occasionally to get the air moving.

✚ Lamb's lettuce doesn't suffer so much from flea beetle, so you can grow it successfully outside, especially since in many areas winters are getting warmer.

✚ Though quick to germinate and grow well, lamb's lettuce is a small plant and risks being outcompeted by quicker-growing weeds, so you need to keep an eye on it in the early stages, and weed early.

✚ Like most salad leaves, lamb's lettuce doesn't like being too dry, so keep well watered.

✚ When it comes to harvesting, lamb's lettuce is a bit small and fiddly so I tend to pick whole plants by cutting off the rosette at soil level. You can get a bit more from them if you can be bothered to pick individual leaves.

ZERO-WASTE TIPS

✚ Leave some plants to go to flower and seed and you will get a nice crop of self-sown plants the following year. Unlike many plants that seed easily, lamb's lettuce rarely becomes a problem as it is so small and doesn't smother other crops.

TOO MUCH LAMB'S LETTUCE?

✚ Though usually used fresh as a salad leaf, lamb's lettuce can be cooked. Steam or wilt it as a spinach alternative, or chop and add to soups. If you have masses of it you can even use it as the main ingredient for a soup – about 500g/1lb for four people, though you will need to add something like chicken or blue cheese to give the soup some depth of flavour.

FENNEL

Foeniculum vulgare var. dulce

I am a sucker for aniseed, so fennel is one of my favourite vegetables. I quite happily munch on the raw bulb or leaves straight from the ground. Despite its tendency to bolt, fennel is still a good contender for zero-waste gardening. Even if the bulb doesn't fully form, you can harvest the leaves and undeveloped bulb or wait till it flowers and develops seeds.

Sow: Late spring to early summer, direct or in modules

Plant: Early summer

Harvest: Late summer to autumn

Eat: Fresh, raw or cooked, or in stocks, soups and stews

Yield

	Per Plant	Per m²/yd²
Total	250g/9oz	8kg/17½lb

GROWING TIPS

✚ Fennel is one of those plants that really doesn't like being transplanted. Though it can work, it increases the risk of bolting. The trick is to sow into largish containers and don't prick out the seedlings. Also make sure the seedlings don't dry out.

✚ If sowing direct then distribute the seed quite thickly (1cm/½in between seeds) and thin the plants out to a final spacing of 20cm/8in. But thin early; if the bulbs get too cramped they might not form proper bulbs. As with most thinnings, you can use these in salads or finely chopped in an omelette.

✚ Like many 'herby' vegetables fennel evolved its strong flavour as a defence mechanism against pests and diseases. Despite our cultivated varieties having a milder taste than their wild ancestors, the mechanism still works and fennel is remarkably pest- and disease-free. It doesn't escape slugs, though, and is vulnerable in the early stages but once it gets going it shouldn't be eaten or infected by much.

ZERO-WASTE TIPS

✚ Most of the fennel plant above ground is edible. Though modern varieties have been bred for a fat, swollen bulb, you can also use the leaves in teas and salads or as an aniseedy herb. The fronds can even be crystalized for a delicious aniseed sweet treat.

✚ The stalks, though a bit tougher than the tender bulb, are still delicious. When they are young, eat them like you would a stick of celery. If they are too tough for your taste then you can cook them to soften.

✚ The flowers, too, are edible, and indeed fennel pollen is a fashionable delicacy with an intense liquorice/citrus flavour. It sells for £1 per gram/US$36 per oz. To harvest, just pick the whole flower head with a bit of stalk. Hang it upside down with a clean paper bag fastened around the flower to catch the pollen. Don't throw the rest of the flower away though; once it has dried you can rub the flower into a powder to use as a flavouring.

✚ Fennel seed too has culinary value, adding a zing to meat dishes particularly (see pages 146–147)

✚ When you cut a fennel bulb, don't dig the plant up but leave it to regrow. It will produce a few little bulblets as a second or even third bonus harvest.

✚ Despite your best efforts a fennel plant doesn't always form a bulb but instead sends up woody green shoots. Are these worth keeping? Yes, these tougher shoots are not worth eating but are a great flavouring in stocks. Just chop roughly and add them in.

TOO MUCH FENNEL?

✚ Pickled fennel is a joy: it keeps its texture well and the strong flavour copes well when pickled, remaining distinctive and interesting.

✚ You can also freeze fennel successfully, though it's best to freeze a young bulb as big ones can end up being tough once thawed. You can still use those for stocks of course.

ORIENTAL BRASSICAS

Brassica rapa

I cannot do justice to the range of crops that come under this banner. They are varied in taste, shape and colour, and are mostly grown for their leaf or tender stems. Many have a peppery or mustard flavour, and some are really fiery, particularly as the leaves get older. As brassicas, they suit the cooler seasons, and do very well in tunnels or greenhouses over winter. There are three main distinct groups of oriental brassica.

Sow:	Spring or late summer
Plant:	Late spring or early autumn
Harvest:	Spring to autumn
Eat:	Fresh; pickled, year-round

PEKINENSIS GROUP

Chinese cabbage and komatsuna belong to this group. They tend to have a compact head like a cabbage. They are great shredded as a lettuce or cabbage replacement, but also work well in stir-fries. Allocate six plants per m²/yd² in two rows of three plants.

Yield

	Per Plant	Per m²/yd²
Total	1kg/2¼lb	6kg/13lb

CHINENSIS GROUP

Many of these have choi/choy in the name: pak choi, joi choi, bok choy for instance. Their structure is more like chard with a thick stem and dark green leaves. They are delicious when steamed, braised or stir-fried, but are lovely raw too, especially when harvested young. Allocate twelve plants per m²/yd² in two or three rows of four plants, or in five rows of ten small plants.

Yield

	Per Plant	Per m²/yd²
Total for large	750g/1½lb	9kg/20lb
Total for small	100g/3½oz	5kg/11lb

NIPPOSINICA GROUP

These are a more leafy, mustardy type, and are usually harvested when young for salads. Though they can still be tender and tasty as they get older, their hotness increases so eat with care.

Yield

	Per Plant	Per m²/yd²
Total	200g/7oz	12kg/26lb
Per pick	50g/2oz	3kg/6½lb

GROWING TIPS

✚ Oriental brassicas are all delicious and tender, and the slugs think so too. Plants are very prone to attack, particularly in the early stages.

✚ The other main pest is the flea beetle, which can destroy a crop if the weather is really hot and dries the seedlings soon after they germinate. Some minor damage from the pest affects only how a plant looks (it leaves little white marks on the leaves) and you'll probably be able to put up with it. You can overcome this by growing in tunnels or greenhouses, or even by covering with a temporary cloche or cover – be aware, though, that you need to get a really good seal at the edges to prevent flea beetles sneaking in under the gaps.

✚ Oriental brassicas are a good crop to grow before potatoes, and can even reduce the risk of wireworm if planting into recently cultivated grass.

ZERO-WASTE TIPS

✚ Pick young. Most oriental brassicas grow really quickly and can become either very hot (in the case of the mustardy ones) or a bit tough. If they do get away you can still eat the flower shoots.

TOO MANY ORIENTAL BRASSICAS?

✚ Big stir-fries.

✚ Oriental brassicas also work really well as pickles, especially Chinese cabbage, which is the traditional ingredient of kimchi.

How often to pick

Pick every few days if harvesting individual leaves. If cutting the whole plant allow 2-4 weeks to regrow. Both will be longer in very cold or very hot conditions

WINTER PURSLANE

Claytonia perfoliata

Sow:	Late spring to early summer, direct or in modules
Plant:	Early summer
Harvest:	Late summer to autumn
Eat:	Fresh, raw or cooked, or in stocks, soups and stews

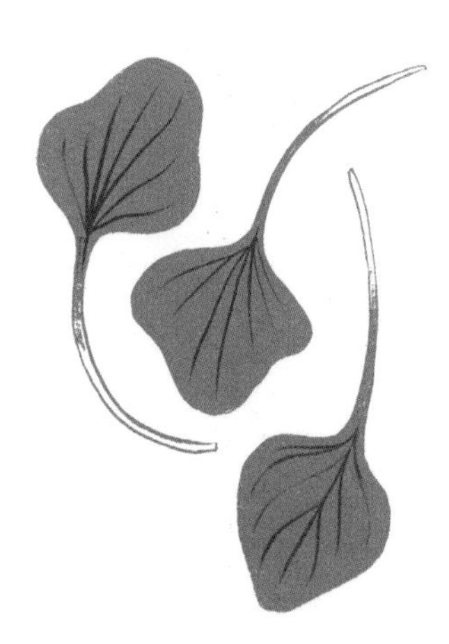

Also called miner's lettuce or claytonia, winter purslane is a great little winter salad. It grows best under cover, though it is super-hardy down to -35°C/–31°F or lower.

GROWING TIPS

✚ Winter purslane germinates quickly and in relatively low temperatures so is useful for interplanting with larger plants or filling gaps later in season.

TOO MUCH WINTER PURSLANE?

✚ Larger leaves and big quantities can be steamed like spinach, or even made into a good soup.

Yield

	Per Plant	Per m2/yd^2
Total	75g (2.7oz)	5.25kg (11.5lb)
Per pick	25g (0.9oz)	1.75kg (3.8lb)

How often to pick

Pick every three to four weeks

SPINACH

Spinacia oleracea

I confess I have mostly given up growing what is called 'true' spinach in favour of the more productive and hardy chards and perpetual spinach. However, there is something wonderful about the more delicate texture and subtle flavour of this crop.

Sow: Early spring or late summer, direct or in modules

Plant: Spring; or late summer to early autumn

Harvest: Early summer; and autumn to winter

Eat: Fresh, summer to early winter; frozen/pickled, year-round

Yield

	Per Plant	Per m^2/yd^2
Total	150g/5oz	4.5kg/10lb
Per pick	75g/2½oz	2.25kg/5lb

How often to pick

Pick every four weeks, though regrowth is dependent on weather

GROWING TIPS

✚ There are three ways to harvest spinach: as a baby leaf for salads; by picking the whole head, although this sacrifices a little yield (about one-third), but this super-quick plant can mostly be used whole without having to strip the leaves from the stems; and by letting the plant grow a bit larger and picking large leaves from the plant on a cut-and-come-again basis. This will get the most out of your plant but is more time-consuming.

✚ Spinach is one of those crops that loves to bolt. In hot or very dry conditions it quickly enters a seed-production phase. Sowing in cooler weather at each end of the season will reduce the risk of bolting, as will making sure you give plants enough water.

✚ The other scourge of spinach is mildew. Though many modern varieties have some resistance, in damper climates it is worth giving your plants a bit of extra space to allow good air movement around the plants.

✚ Based on two picks 20cm/8in spacing three rows per m sq (9ft sq)

ZERO-WASTE TIPS

✚ The best way to reduce waste with spinach is not to let it go off in the fridge. I prefer to eat it within a day or two – as with all leafy veg the fresher you can eat it the better. If I do need to keep spinach I wash it carefully to avoid bruising, dry carefully and pack loosely into a reusable plastic bag, tub or glass jar.

TOO MUCH SPINACH?

✚ Lots of fresh leaves cook down to not very much at all, so I rarely find myself with too much spinach, but if there is a glut you can add it as a cooked ingredient to pasta dishes like lasagne, or to quiches. Spinach goes well with cheese, for instance in dishes like spanakopita.

✚ Spinach freezes well. If you don't want to keep it for long you can just wash, spin dry, then pack the leaves into a tub or bag and freeze. However if you want spinach to keep well for longer then you'll need to blanch it first.

✚ Spinach can work as a pickle too but it doesn't have a strong flavour so is good mixed with other ingredients and lots of spices.

CHARD

Beta vulgaris subsp. *vulgaris*

Chard and its relative 'perpetual spinach' are popular vegetables with professional growers. They are productive, hardy and not too prone to pests and diseases. For home gardeners they offer the same potential, and look stunning too. I have often included some plants in my ornamental borders.

Sow:	Spring or late summer (or even early autumn under cover)
Plant:	Spring to early summer; or late summer to early autumn
Harvest:	Most of the year round, unless heavily frozen for long periods
Eat:	Fresh, or frozen

Yield

	Per Plant	Per m^2/yd^2
Total	1.3kg/3lb	26kg/60lb
Per pick	325g/0.75lb	6.5kg/15lb

How often to pick

Pick every week or two if harvesting individual leaves, or two to four weeks if cutting the whole plant.

GROWING TIPS

✚ Along with kale, chard is one of the easiest ways to produce fresh leaves for the kitchen. Swiss chard has the biggest leaves and thick white succulent stems, while ruby and rainbow chard are slightly smaller but with vibrant colours that are mostly retained during cooking.

✚ Perpetual spinach is less visually stunning but even hardier, and easier to grow; as the name suggests it keeps coming, providing a long harvest of leaves.

✚ Space chard plants at about 30cm/12in apart in rows 45cm/18in apart.

✚ You can harvest in two ways: pick individual leaves from the outside, leaving younger leaves to develop and grow; or just cut the whole plant down to a few centimetres leaving the plant to regrow. The latter technique is quicker, but you end up with a range of different-sized leaves, and some of the smaller ones end up with cut edges. There is also the risk of bits of dead leaves getting into the crown so I prefer to harvest individual leaves even if it takes a little bit longer.

ZERO-WASTE TIPS

✚ Chard, and particularly Swiss chard with its huge juicy stems, is effectively two vegetables in one. The leaf is spinach, and can be cooked in the same way, though it holds its flavour and structure better than true spinach when cooked. The stem is separate and needs to be treated differently: I usually finely chop the stems and fry them for 5–10 minutes before adding in the sliced leaves for another couple of minutes. This retains some crunch. Alternatively, for a softer deeper flavour you can add some wine or vinegar, cover and cook very slowly on a low heat for another 30 minutes.

✚ Swiss chard has been developed from sea beet, and you can tell from its leaves that it is the same family as beetroot, so it's no surprise that chard root is edible, even if it hasn't been bred specially for that purpose. However, if you have been harvesting leaves from your plants for months then the roots are likely to be tougher than a lovely young beetroot. When roasted long and slow the root can still be edible, and has a rich nutty flavour.

✚ If you've planted more chard than you need and can't keep up with the leaf harvest then it's worth pulling a few plants up and using the roots while they are still relatively young and tender.

✚ I sometimes use the really big leaves as savoury wraps. Carefully steam them whole and fill with something rich and creamy. I usually use a couple at a time to make sure they don't collapse.

TOO MUCH CHARD?

✚ Prepare the stems and leaves separately but both can be blanched and frozen in bags.

✚ Chard stems are great pickled, and if you have a range of coloured stems they will look stunning in their jar. Cut the stems very small to use in burgers for instance, or leave bigger to eat as an aperitif.

LETTUCE

Lactuca sativa

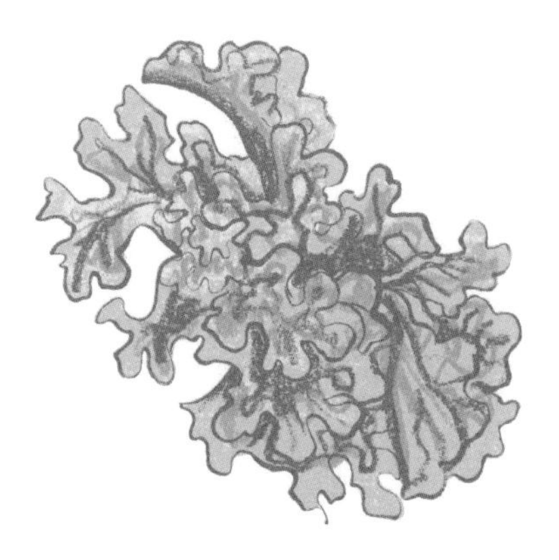

When I was growing up in 1970s England you could buy three types of lettuce: traditional cos; butterhead; and the modern fashionable, though fairly tasteless, iceberg. Now there is a much greater range available to buy, and when you grow your own lettuce the choice is even wider.

Sow:	Spring to autumn
Plant:	Spring to autumn
Harvest:	Spring to autumn
Eat:	Fresh, spring to autumn; preserved, year-round

How often to pick

Pick every few days if harvesting individual leaves. If cutting the whole plant allow 2–4 weeks to regrow.

Yield (whole head)

	Per Plant	Per m²/yd²
Total for large	800g/1¾lb	7.25kg/16lb
Total for small	300g/10oz	6kg/13lb

Yield (loose leaf – cut-and-come-again)

	Per Plant	Per m²/yd²
Total	150g/5oz	7.5kg/16½lb
Per pick	25g/1oz	1.5kg/3¼lb

GROWING TIPS

✚ The broadest lettuce distinction is between cut-and-come-again varieties, and those that form a whole head (though some work well either way). Cut-and-come-again are usually sown directly and more densely, while headed lettuce works well when sown in seed trays or modules and transplanted to a final spacing.

✚ Salanova types are a relatively new group with striking colours that look like a whole head but with one cut become a mass of small leaves.

✚ Lettuces take 45–100 days to mature, though you can harvest them at any point if not worried about getting maximum yield. To get a continuous supply of lettuces you'll have to sow continuously.

✚ Commercial growers often sow every week or two, but as gardeners we can leave it a bit longer than that. I would suggest sowing a few plants or rows every three or four weeks.

Now there is a much great range of lettuce available to buy in the shops, but when you grow your own the choice is even wider.

✚ Lettuce seed won't germinate if it's too hot so if possible find a cool place to sow your lettuce, or do in the evening so it at least has the cool night to start.

✚ The weather will affect maturity too: lettuce likes it cool with regular water and tends to stop growing if it is very dry or hot.

✚ Some varieties have been bred especially for low light levels and cold winters. Though not always outside hardy, they can do well in tunnels and greenhouses. Varieties such as 'Winter Marvel' and 'Reine de Glace' can even be sown in late autumn, and again in midwinter under cover.

✚ Interestingly 'Reine de Glace' is also fairly heat-tolerant and is slow to bolt in summer. There are other varieties that will fare better with hot summers. Choosing a range of varieties will give you the best chance of success.

ZERO-WASTE TIPS

✚ Lettuce seed is very easy to save, unless it rots in a cool damp summer, so if some of your lettuce does flower don't pull it all out but leave a few plants. Lettuce is mostly self-pollinating, so home-saved seed will generally give you the same variety as you sowed, and often germinates better than bought seed.

✚ Even if you are not growing to cut-and-come-again, if you cut your lettuce head high (2.5cm/1in or so above the ground) you should get some reshoots for a secondary crop.

TOO MUCH LETTUCE?

✚ Lettuce is usually eaten raw in salads, but can be cooked very happily. Lettuce soup is delicious, or add the leaves to stir-fries, quiches or braise them whole.

✚ Those leaves that don't have a thick stem are great for stuffing, as you would a cabbage leaf, though double them up as they are not as strong as cabbages.

✚ Lettuce pickles well either as whole leaves if they are not too big, or when ripped up smaller. They are delicious flavoured with dill.

WATERCRESS

Nasturtium officinale

Sow:	Spring or late summer
Plant:	Spring to early summer; or early autumn
Harvest:	Summer to autumn
Eat:	Fresh, or in soup

Since most of us don't have a fresh spring in our garden I've concentrated on growing watercress as a quick annual in pots or in the garden. It works well, though you need to keep plants well watered.

GROWING TIPS

✚ Treat watercress like a salad crop: sow a few seeds into a pot or module and then transplant into a larger pot or the ground for growing on.

✚ Watercress needs moist soil. It will grow well in a pond with a pump, but it's vital to keep the water really clean to avoid contaminants and potential infection.

Yield

	Per Plant	Per m^2/yd^2
Total	1.3kg/3lb	26kg/60lb
Per pick	325g/0.75lb	6.5kg/15lb

RHUBARB

Rheum × hybridum

I can do no better than to quote Lawrence D. Hills: '[Rhubarb] comes before the gooseberry dares, as the first "fruit" of every spring. Bird proof, frost proof, town proof, it is nutritionally superior to plums and greengages and for the least trouble of almost anything that grows in the garden.'

Sow: Spring
Plant: Autumn or winter
Harvest: Spring to early summer
Eat: Fresh, spring and early summer; preserved/frozen, year-round

Yield

	Per Plant	Per m^2/yd^2
Total	2kg/4½lb	2kg/4½lb
Per pick	400g/14oz	400g/14oz

How often to pick

Pick every few days, or as needed.
PIcking encourages young tender growth

GROWING TIPS

✚ Rhubarb has two big things going for it: it is very hardy; and is pest- and disease-proof so great for the zero-waste gardener. It also starts producing right at the start of the hungry gap – that period at the end of spring when winter crops have finished and summer ones have not yet started. The fat rhubarb buds bursting through the soil are a sign that the dark days are on their way out and that fresh vegetables are on their way in.

✚ Although you can buy rhubarb seed, there is only a small range of varieties available and they are not the sweetest so I would recommend buying crowns for planting. You should be able to crop the plant for at least ten years, so the cost is worth it.

✚ At some point your plant will become less productive, that is the time to divide it. This means you should dig it up and split up the crown into three or four pieces (depending on the size of the plant). Replant the pieces to grow on for another ten years or more.

✚ Rhubarb is one of the few crops that really needs a cold winter in most locations. It becomes dormant when day length shortens to less than ten hours a day, and this dormancy needs to be broken by a period with temperatures lower than 4°C/39°F. If you are growing rhubarb in the tropics it won't become dormant.

✚ Forced rhubarb is a term given to excluding light from the growing shoots to yield a pale pink and sweeter stem. Often the plants are kept in warmer conditions to encourage quick growth, though you can do it by just placing a big bin over the plant. Don't force the same plant every year though or it will become weak. I force one-third of my plants every year and then give them a couple of years to recover.

ZERO-WASTE TIPS

✚ Rhubarb has big leaves that are not edible – indeed they contain high levels of oxalic acid, which make them poisonous. However that doesn't mean they aren't useful. I use rhubarb leaves for weed control, placing them around perennial plants. They don't last long but do help to slow weed growth.

✚ Rhubarb leaves can even be used as a way to clean your burnt pans. Just boil up a few leaves with some water and the acids in the leaves will help remove those burnt stains.

TOO MUCH RHUBARB?

✚ Jam and chutney are the obvious ways to preserve rhubarb. It is also very easy to freeze – just cut into chunks and freeze on a tray before bagging up.

✚ The tartness of rhubarb also lends itself to making syrup, which is delicious on ice cream or pancakes.

✚ For larger quantities you can make rhubarb wine, which mixes well with sweeter fruit. Freeze the rhubarb at harvest and then thaw it to mix with the other fruit. The freezing helps to break down the fruit and start the fermentation process.

BRASSICAS

Brassica oleracea

All of the crops we grow in the brassica family are very closely related, and indeed will cross-pollinate readily. There are two main species: *Brassica oleracea* (which contains all the cabbages, kales, sprouts, cauliflower and calabrese) and *Brassica rapa* (turnips, mustards and oriental brassicas). Humans have selected different brassica types for particular traits, like flower, leaf size or colour, and ability to heart up. However, aside from those special attributes, how we can best make use of the whole plant when growing and eating brassicas is similar for all. We will go into the separate brassica crops in more detail later, but here is some general guidance on growing and harvesting them as a group.

GROWING

✚ As you might guess from looking at their big green leaves and large plant size, brassicas are heavy feeders. You want strong plants that are not susceptible to pests. They need a rich soil with plenty of fertility, so add compost or well-rotted manure in the part of the rotation before these crops. However, if there is too much nitrogen then, like many plants, brassicas will grow too quickly and have thinner cell walls, which makes them easy prey for sap-sucking insects and fungal diseases. You want them to grow strongly but not too quickly. Adding your compost or manure to a growing cover crop in the previous part of the rotation reduces the risk of excess soluble nitrogen.

✚ Brassicas are particularly suitable for cooler climates. They germinate in low temperatures and grow well even at temperatures of 10°C/50°F and in lower light levels. They suffer in very hot and dry conditions. For this reason they are often thought of as a winter crop to be sown early and late in the season to provide nutrient-rich, fresh food at times when other crops are less available.

✚ If growing strongly brassicas are fairly pest- and disease-free. However birds – particularly pigeons – love them, particularly if it snows and their tall stalks are the only green thing left visible poking above the white. Covering with a mesh is the most effective way to keep birds off, but this doesn't quite fit with our zero-waste approach. I have had some success with stringing CDs through the crop, as they blow in the wind and catch the light they spook the birds . . . but only for a while.

One superpower that brassicas have is that every part of the plant apart from the root is edible, though of course some parts are more tasty than others

EATING

✚ One superpower that brassicas have is that every part of the plant apart from the root is edible, though of course some parts are more tasty than others. The leaves from all plants can be harvested, even if that is not the main crop you have grown them for. For example Brussels sprout tops were a classic crop historically in the UK, and the leaves around a cauliflower are also delicious to eat.

✚ For those crops that produce a head (cabbages and cauliflower), once it is harvested the plant will often produce a second crop of mini-shoots that are superb in stir-fries or steamed.

✚ **Flowers** All brassicas tend to go to flower and seed, usually after you have harvested the main parts you have grown them for. Harvest the young yellow flowers for eating in salads or steamed.

✚ **Seed pods** Technically all brassica seed pods are edible;

however unless you catch them really young they're pretty tough and unappetizing. Cooking them will help soften them a bit, however apart from some radish varieties that have been specifically bred for edible seed pods, I wouldn't particularly recommend them.

✚ **Seeds** Brassica seeds can be eaten but are probably best sprouted first. Alternatively save the seed and use it as a quick green manure cover or baby leaf crop. You can let these self-saved plants grow bigger, but brassica plants are notoriously promiscuous and they all cross with each other (and even with their wild relatives) so you will end up with a mixed crop of probably mostly kale-like plants. It is always fun to try though.

PRESERVING

✚ Though frozen brassicas are edible, I find they become mushy and less appetizing. With the exception perhaps of soups, pickling is a better way of preserving most brassicas when you have a glut.

Don't be too tidy

As an advocate of messy gardening, I recommend not cutting down all your brassicas when they finish cropping. Leave some to flower over winter. As well as being a good source of food for a range of wildlife, they provide a habitat for the parasitic wasps that attack cabbage white butterflies. If you can keep a good population of these wasps you will never suffer serious damage from cabbage whites again.

BRUSSELS SPROUT

Brassica oleracea Gemmifera Group

Sow:	Early spring
Plant:	Late spring to early summer
Harvest:	Winter
Eat:	Fresh

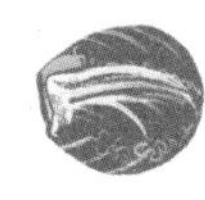

Though not the world's favourite vegetable, I love sprouts. I usually cook them without water by frying up with some bacon and garlic. It's a bit tricky to grow a good even crop, but if you like them they are well worth the effort.

GROWING TIPS

✚ If you cut off the sprout top in early autumn, the sprouts will be encouraged to grow evenly up the stem. This was a standard technique among market gardeners to increase the value of the crop and improve saleability of sprouts on the stalk.

ZERO-WASTE TIPS

✚ Don't forget to eat the sprout tops, which grow like a very loose-leaved cabbage.

Yield

	Per Plant	Per m²/yd²
Total	1kg/2¼lb	3kg/6½lb
Per pick	250g/9oz	750g/1½lb

BROCCOLI

Brassica oleracea Cymosa Sprouting broccoli
Brassica oleracea Italica Calabrese

It can all get a bit confusing here, with many people calling calabrese 'broccoli'. Calabrese forms a single big head, while the sprouting broccolis produce a large number of smaller sprouts. The speciality Romanesco and new cauliflower/broccoli hybrids, can be confusing too. I have grouped all the brassicas (apart from cauliflower) for which we eat the flower sprouts here.

Sow: Early spring (calabrese), late spring (sprouting broccoli)

Plant: Late spring (calabrese), early summer (sprouting broccoli)

Harvest: Summer to autumn (calabrese), winter to spring (sprouting broccoli)

Eat: Fresh

Yield

	Per Plant	Per m^2/yd^2
Total	500g/1lb	1.75kg/4lb
Per pick	150g/5oz	600g/1¼lb

GROWING TIPS

✚ You can crop for most of the year by carefully choosing a range of varieties that have been bred to extend the cropping season, though exact timing will depend on local and annual climatic conditions. For instance, many of the sprouting broccolis come into flower after a cold winter. If you don't get those frosts they may not develop, or even if they do they may not taste quite so good. Don't despair though; there are varieties available that don't need this 'vernalization' period.

✚ There are also different colours and shapes, like the extraordinary looking Romanesco or white sprouting broccoli. Some can be tricky to cultivate so start with a tried-and-tested sprouting broccoli such as 'Early Purple Sprouting' and then experiment with others to see which suits your taste and growing site.

✚ Sprouting broccoli can remain in the ground for almost a year, so making sure it has the best start is crucial. Like other tall brassica plants, it helps to firm them in well when planting to avoid windrock. Brassica plants will root from their stems, so don't be afraid of planting deep – though don't cover the growing tip. The additional rooting area will help produce a solid plant.

✚ As it has such a long season to develop and is spaced widely it's perfectly suited to interplanting with a quick crop (see page 24.) or undersowing with a fertility-building green manure crop.

✚ All broccoli plants produce a big central head, and then subsequent smaller shoots. With calabrese the central shoot is the main harvest, with a few subsequent bonus flower shoots on a secondary harvest. For sprouting broccoli this is reversed: the central head is relatively small and the side shoots are the main harvest and keep developing the more you cut, though they will get gradually smaller. They are best harvested when the buds are as tight as possible, though you can eat them even once the flowers have opened.

✚ Plant spacing will affect how the flowers are produced – wider spacing encourages more side shoots. For sprouting broccoli this can be the most effective use of space, particularly if you interplant as it grows.

ZERO-WASTE TIPS

✚ Though my kids won't eat them, I prefer the stems to the flowers. Cut the stem long when harvesting to make the most of them. Peel the stem as you get further from the flower or it will be tough.

✚ If they are not going to be eaten immediately, the best way to keep calabrese heads for as long as possible is in a glass of water (like a cut flower) in the fridge.

TOO MUCH BROCCOLI?

✚ I have never had too much sprouting broccoli. It comes at a time of year when fresh greens are scarce and I can eat piles of it simply drizzled with butter, and seasoned with salt and pepper.

✚ Calabrese is a different matter, with a milder flavour. Coming as it does when other vegetables are in full flow, you might well find yourself faced with a glut. I love it grilled or roasted either hot or cooled in a salad.

✚ Broccoli stems pickle beautifully, so eat the flower heads fresh and pickle the stems for later.

CAULIFLOWER

Brassica oleracea Botrytis Group

In my twelve years as a professional vegetable grower I never grew a totally successful cauliflower crop, though I had some nice individual specimens. So if you have plenty of space and like to challenge yourself then give them a go, but otherwise you might decide to concentrate on other crops.

Sow: 6–9 months before predicted harvest date

Plant: When seedlings have 3–4 true leaves

Harvest: When curd is white and tight

Eat: Fresh, or pickled

Yield

	Per Plant	Per m^2/yd^2
Total for large head	850g/1.9lb	2.5kg/5.5lb
Total for small head	300g/10oz	2.75kg/6lb

GROWING TIPS

✚ Timing is the key. Varieties have been bred to give almost year-round production in most climates. The harvest date will determine sowing date, though planning an exact harvest time is tricky, particular with the UK's increasingly unpredictable weather patterns. Varieties that should mature in a convenient sequence can end up all coming at once (or not maturing) if we get a long period of unseasonal hot or cold temperatures.

✚ Grow open-pollinated varieties as they will be less likely to all mature at once. Unlike professional growers who want lots ready at the same time so they can supply full boxes to customers, as home gardeners we want one every few days if possible. The genetic diversity in open-pollinated varieties should help spread the harvest.

✚ There are an increasing number of coloured cauliflower varieties available: purple, orange and yellow. Though I have not grown them all I have generally found them to be less vigorous and productive than the traditional white ones.

✚ I spoke to a seed breeder a few years ago who had bred a green variety that was supposedly the hardiest, most pest- and disease-resistant, most flavourful cauliflower they had produced, but they couldn't sell it because people wanted white curds. Since then there have been some green varieties released, with one company rebranding them broccoflower to try to overcome the conservatism of consumers.

Harvesting tip

The head we eat is actually a cluster of flower buds that don't hang around long before bursting into flower. Hence there is a relatively small window to harvest cauliflowers between 'fantastic it's big enough' and 'oh no, it's gone over'.

ZERO-WASTE TIPS

✚ Don't throw away the leaves. Though you may need to trim off the brown edges where they have been cut, cauliflower leaves are delicious. They have a thick stalk that needs more cooking than cabbage, so I tend to separate the stalk and leaf and then thinly slice the stalk, which can be steamed or braised, adding the leaf at the end for a couple of minutes' cooking.

✚ As with cabbage, retain enough stalk when harvesting to get some regrowth from the plant in order to have a secondary crop.

TOO MANY CAULIFLOWERS?

✚ Since cauliflowers have a tendency to mature at the same time, it is likely that you will have a glut at some point. There is only so much cauliflower cheese you can eat.

✚ Cauliflower pickles remain brilliantly crunchy. They look great in the jar too, especially if you add some brightly coloured peppers and chillies.

KALE NERO

Brassica oleracea Acephala Group

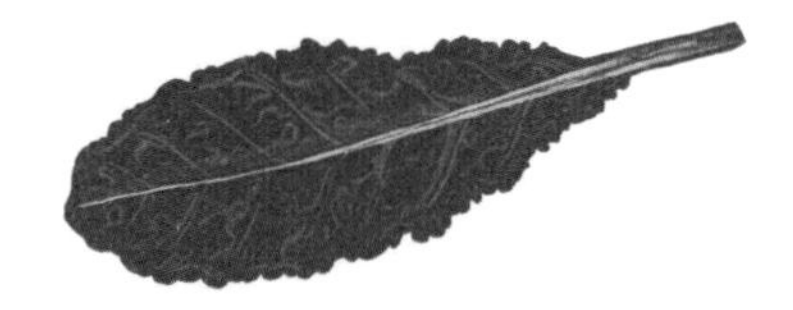

Kale is one of life's survivors, being able to withstand even the brutal climate of islands like Shetland in the middle of the North Sea. Though historically a major source of vitamins in northern climates, kale lost popularity when we could import more diverse vegetables during winter. It is now once again celebrated, and new varieties are being bred.

Sow:	Early spring
Plant:	Spring
Harvest:	Summer to winter
Eat:	Fresh

Yield

	Per Plant	Per m^2/yd^2
Total	900g/2lb	3.6kg/8lb
Per pick	225g/8oz	900g/2lb

How often to pick

Pick every two to three weeks.
Growth slows in very cold weather.

GROWING TIPS

✚ If I had to choose just one brassica to grow, it would be kale. Given half-decent soil, and provided you don't live in a desert, you should be able to get a crop in most years. It doesn't suffer much from pests and diseases, and since you just pick the leaves it doesn't need critical periods of cold or rain like some of its cousins do to form a head or bulb.

✚ There are a range of varieties from the rugged Shetland kale, through to the more delicate Italian black kales.

✚ Kale grows particularly well from bare-root transplants (see page 44.) but are also easy to grow in pots or modules.

✚ When harvesting I tend to pick a few medium-sized leaves from each plant, working my way up as the plant grows bigger. You can also cut the head out – it will look almost like a small kale plant – which will encourage more side growth of smaller leaves.

ZERO-WASTE TIPS

✚ The beauty of kale is that it can be harvested at pretty much any stage of growth: you can grow as a baby leaf salad crop; let it grow on to 15–20cm/6–8in tall and cut as a whole plant; or take the traditional route and let it grow to full maturity.

✚ Kale is grown for its leaf greens while the leaf stem often gets consigned to the compost heap. However even those stems that seem big and tough can be delicious if cooked right. Here are a couple of ways to try. In both cases blanching the stems first will help soften them and reduce toughness.

✚ Slice the stems thinly, fry with a little butter or oil over a low heat until tender. You can add garlic or herbs for flavour.

✚ Cut the stems into short lengths (2.5cm/1in or so), dust with flour and deep-fry.

TOO MUCH KALE?

✚ Kale crisps (chips) are a great way of using up a lot of kale. Tear up the leaves into smallish pieces and, once washed and dried, rub some oil into them, then cook in a low oven (150°C/300°F) for about 20 minutes. Add spices as well as salt and pepper for flavour.

✚ Alternatively, add kale to almost any dish as a spinach substitute, to food like quiches and omelettes, or even as a pesto ingredient.

CABBAGE

Brassica oleracea Capitata Group

There's a cabbage for every season, from the loose green head of spring greens to the tight red cannon ball in winter. Each takes a bit of space and is in the ground for a while so it is not always the first choice for those with very small gardens, but cabbage is worth the effort if space is not an issue.

Sow:	Early spring (summer cabbage), late spring (winter cabbage), late summer (spring greens)
Plant:	When plant is big enough: 7cm/3in high
Harvest:	When head is formed, but before it bolts or splits
Eat:	Fresh, or pickled

Yield

	Per Plant	Per m²/yd²
Total for large head	1kg/2¼lb	4.5kg/10lb
Total for small head	700g/1½lb	2.75kg/6lb

GROWING TIPS

✚ Most cabbage varieties are hardy and relatively easy to grow; however they do take time to grow and the longer something is in the ground the more chance there is that something will happen to it.

✚ Spring greens are fantastic as they provide fresh leaves during the hungry gap in late spring and early summer, but they need to be nursed through the winter as smallish plants. Growing them under cover will help. You can use cloches if you don't have a tunnel or greenhouse.

✚ Cabbages have been bred as cold-climate crops, with some varieties specifically designed to withstand long periods in below-freezing temperatures, like 'Mid-Winter King'. With our increasingly warm and wet winters, though, our cabbage plants are at risk from higher slug populations in winter. Once slugs get into the heads they can cause lots of damage without us seeing. You can reduce this risk a little by removing some of the lower leaves to keep the plant a bit drier underneath.

ZERO-WASTE TIPS

✚ When harvesting cabbage, leave 5cm/2in of stalk in the ground. This will form a little cluster of new shoots that can be harvested and are a tasty bonus.

TOO MANY CABBAGES?

✚ All cabbages pickle well, but I particularly love the look of red cabbage in the jar.

✚ How about a big pile of bubble and squeak? Traditionally this is made from leftovers, and is great if you cook the cabbage the day before using it for bubble and squeak. Mix up cabbage cooked in meat fat or oil, with mashed potato and some fried bacon. I can eat piles of it!

KOHL RABI

Brassica oleracea Gongylodes Group

This vegetable is not widely known, but it is actually one of my favourites. It has a juicy crisp texture a bit like an apple, but with the taste of a broccoli stalk. It is not hard to grow and can produce an early crop just at the end of the hungry gap.

Sow: Early spring for summer crop; mid-to late summer for an autumn crop

Plant: Late spring for summer crop; late summer for autumn crop

Harvest: When 'bulb' is the size of a small fist

Eat: Fresh or pickled

Yield

	Per Plant	Per m²/yd²
Total	150g/5oz	3kg/6½lb

GROWING TIPS

✚ The key to a good kohl rabi 'bulb' – actually a swollen stem – is regular and even watering. If it has periods without water the stem will get tough. The better you can look after it the bigger you will be able to let it grow without it becoming fibrous.

ZERO-WASTE TIPS

✚ Kohl rabi leaves are delicious, but dry out quickly so need to be eaten soon after harvest.

✚ If harvested young, you will be able to eat the skin; older specimens, even where the flesh is tender, will have a tough skin that it is better to peel off.

TOO MUCH KOHL RABI?

✚ It's delicious when grated into a salad or coleslaw, or when sliced thinly in a stir-fry.

This vegetable is not widely known, but it is actually one of my favourites. It has a juicy crisp texture a bit like an apple, but with the taste of a broccoli stalk.

BEETROOT

Beta vulgaris

Often thought of as a winter vegetable, beetroot is one of the easier crops to grow. I also love beetroot as a colourful and meaty addition to summer salads: for instance when grated raw in coleslaw or when gently roasted and served with goat's cheese and walnuts.

Sow:	Spring or late summer, direct or in modules
Plant:	Late spring or late summer
Harvest:	Summer to winter
Eat:	Fresh, summer and autumn; from store, winter; pickled/frozen, year-round

Yield

	Per Plant	Per m^2/yd^2
Total	150g/5oz	4.5kg/10lb

GROWING TIPS

✚ Like many root crops, beetroot doesn't really like having its roots disturbed during growing. It also grows quite quickly so is often direct sown.

✚ However, if you are careful when transplanting them, beetroot can also work well when sown in modules.

✚ Each beetroot 'seed' actually contains 2–5 seeds, and though not all of them will germinate you need to remember this when working out spacing.

✚ One technique that works well with beetroot is multi-sown modules. Put two or three seeds in each module – this will likely give 4–8 beetroot plants. Leave more space when transplanting (30cm/12in rather than 15cm/6in). You can thin out a few babies as they develop (eating them of course) leaving four beetroot max per station to grow to maturity.

✚ If sowing directly I tend to plant densely and thin to a final spacing. This makes the best use of space, providing an early crop of baby beetroot in the hungry gap if weather is warm enough. One seed every 5cm/2in can work well. If you don't thin, the plants will compete with each other and will probably bolt.

✚ Though not totally immune to damage, their slightly bitter taste means that beetroot leaves are not number one on the slug menu. Growing beetroot in modules so the leaves are slightly bigger by the time of planting reduces the risk of them being munched even further.

ZERO-WASTE TIPS

✚ Beetroot is the same family as chard and the leaves can be used in exactly the same way. Cut and cook the leaves immediately even if you are not planning on cooking the root till later.

✚ If you don't manage to harvest all your beetroot, let some go to seed and save it. Sow thickly and cut once as a quick baby leaf pre-crop before late-planted crops like squash or runner beans. The baby leaf of beetroot is lovely in a salad, though you need to pick very young before it gets tough. Beetroot pollen can travel kilometres, so if you save seed you may well not end up with the same variety it came from, but if harvested as leaf then this will not matter.

TOO MUCH BEETROOT?

✚ Pickled beetroot is a classic, but you can also store beetroot well in a clamp (see page 41), or even in the ground with a covering of straw over the top. It will suffer in very hard, prolonged periods of sub-zero temperatures.

✚ Beetroot freezes well, having been blanched or roasted before freezing.

CELERIAC

Apium graveolens var. rapaceum

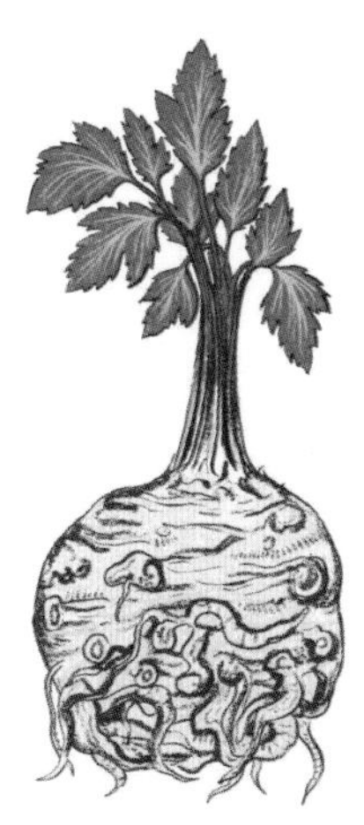

I've never understood why celeriac is not grown and eaten more. It is one of my favourite vegetables. It is a bit of a vegetable diva and needs some coaxing to make a great crop but the hard work is so worth it when it comes good.

Sow:	Early spring, in a seed tray or modules
Plant:	Late spring 30cm/12in spacing in rows 45cm/18in apart
Harvest:	Autumn to winter
Eat:	Fresh, autumn to winter; from store, winter; pickled/frozen, year-round

Yield

	Per Plant	Per m^2/yd^2
Total	150g/5oz	4.5kg/10lb

GROWING TIPS

✚ Celeriac hates drying out, especially in the early stages. Even a short period of water stress can cause problems later on with early bolting.

✚ The seed is tiny and needs light to germinate, so when sowing it's important not to cover with too much compost. Either put the slightest sprinkle of compost over the seed or just scatter the seed on the surface of the compost and firm the seed in gently.

✚ You can sow in modules, or in seed trays, and prick out the best seedlings. If you do sow in seed trays, be very careful not to damage the roots or stem – celeriac plants are still very small when they are transplanted and it is fiddly. I tend to sow it in modules is to avoid the delicate pricking-out operation.

✚ Germination is sometimes patchy and growth slow, so don't get disheartened if it seems like not much is happening. The tricky part is keeping the compost moist enough but not letting it get so wet that the seeds damp off. Some composts appear to have dried at the surface but are still holding plenty of moisture underneath.

✚ Once planted out celeriac still doesn't like drying out. Mulching will help, especially if you are on a light sandy soil. Celeriac has a shallow rooting system so is vulnerable in hot summers; you may need to water regularly to ensure good growth and full bulb.

✚ You can try tighter spacing when planting out to give more but smaller roots per area; however if they start competing too much there is a risk of bolting.

ZERO-WASTE TIPS

✚ Celeriac leaves are a bit tough to eat raw but can be slowly cooked and have a rich celery-type flavour. Towards the end of season, when the leaves are getting older and tatty, I tend to use them for stocks and soups instead.

✚ Celeriac is normally peeled and the peel thrown out, but apart from the really rooty bottom end, which is very fiddly to clean, you can actually use a lot of it. Shred or thinly slice the peel and add to sauces and rice dishes.

TOO MUCH CELERIAC?

✚ You can eat celeriac raw, though it sometimes causes indigestion. My favourite way to prepare it is braised for a long time in olive oil, wine and garlic. However if you need to use up lots then soup is the answer: in cream of celeriac I use double cream but it's great with coconut milk too.

✚ Grating celeriac into risotto at up to 50 per cent volume is another great way to enjoy gluts.

CARROT

Daucus carota

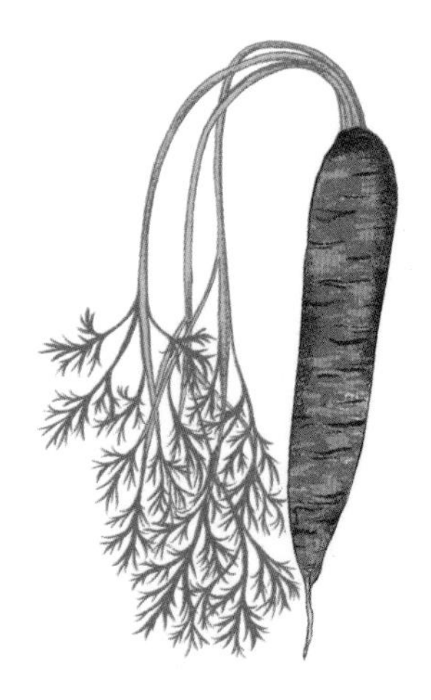

One of the most popular vegetables, happily carrot is also the one that my children have never randomly decided not to like any more. It's a good one to grow, although the slugs like the seedlings. Once it gets going it grows quickly and should be relatively pest- and disease-free – apart from the dreaded carrot fly.

Sow:	Spring to late summer, direct
Plant:	n/a
Harvest:	Summer to winter
Eat:	Fresh, summer to winter; stored/pickled/frozen, winter to spring

Yield

	Per Plant	Per m^2/yd^2
Total	50g/2oz	6kg/13lb

GROWING TIPS

✚ Being such a popular crop there are literally hundreds of carrot varieties to choose from; the majority are now orange but were originally yellow and purple. These early colours are now becoming popular again, along with some white varieties. You can also choose different carrot shapes and those suitable for growing at different times of year. As a result of all these factors, predicting yield is not easy, but expect 100 medium-sized carrot plants per m^2/yd^2, and it is reasonable you might get up to 175 per m^2/yd^2 if you grow them closer and smaller.

✚ Carrots won't stand being transplanted so need to be sown directly into a good seedbed. If you have heavy clay soil, which is tricky to cultivate into a good tilth, you might like to try this good trick. Create your seed drill as best you can, sprinkle the seed into it, then cover not with your soil but with a thin layer of fine bagged compost. This should give good germination; once the plants get going they should be fine even in heavy soil. On really heavy soil choose shorter-rooted varieties; even though you might get some odd-shaped roots they'll be delicious.

✚ Carrot flies are one of the biggest problems for carrot growers. They infect the root and burrow, which causes black discoloration and a bitter taste. The most effective way to stop them is to put a mesh over the crop to prevent the flies coming in and laying their eggs. However, if you are trying to avoid using plastic there are a couple of things you can do to reduce the risk instead. One is to grow only early-season carrots, which are harvested before the larva develop. Or you can plant other strong-smelling plants like onions around your carrots: this helps to confuse the adult flies, which find the crop by smell. Finally, when you thin your seedlings, try to handle them very gently to avoid bruising them and releasing the smell from the leaf, and don't leave the thinnings next to the crop – again the smell from the thinned leaves will attract the adult flies.

ZERO-WASTE TIPS

✚ First off, yes you can eat the tops. If they are young and really fresh you can eat them raw; once they get a bit older they're better cooked. Chop finely as a parsley alternative or add to stocks and stews instead of (or as well as) celery.

✚ A nice treat in winter if you have some big carrots with gnarly tops is to cut the top 2.5cm/1in off and stick it into a pot of compost (or even a tray of water). The resulting sprouts are a lovely source of fresh green flavour.

TOO MANY CARROTS?

✚ Whether pickled, dried, frozen or stored in a clamp (see page 41), there are so many ways to store carrots that you should never need to waste them.

✚ They are also great in juices and soups.

POTATO

Solanum tuberosum

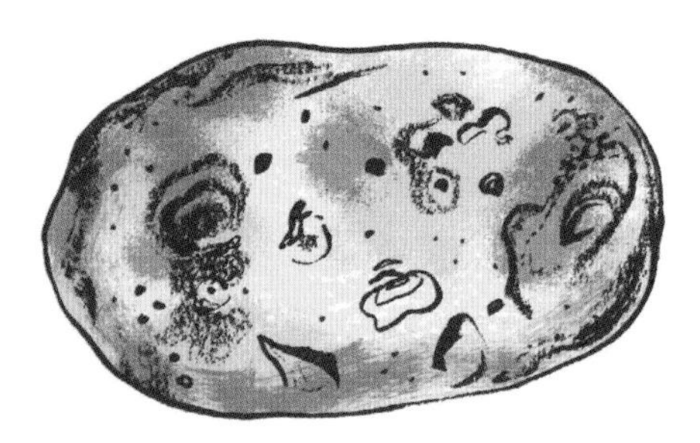

Though easy to grow, potatoes do take up a lot of space and are relatively cheap to buy. Unless you have loads of space I'd concentrate on a few early new potatoes and perhaps some unusual ones that you can't readily buy from the shops.

Sow:	Early spring (early varieties), mid-spring (other varieties) 40cm/16in spacing in rows 75cm/30in apart (slightly closer for earlies)
Plant:	n/a
Harvest:	Mid- to late summer
Eat:	Fresh, late summer; from store, autumn to winter

Yield

	Per Plant	Per m²/yd²
Total	325g/11.5oz	2kg/4½lb

Potato type	Time to maturity
First early	10 weeks
Second early	13 weeks
Early maincrop	15 weeks
Maincrop	20 weeks

GROWING TIPS

✚ Potato growing is very well suited to mechanization. This means that professional growers are able to grow potatoes very well and very cheaply. They also store and travel well. So, although they are easy to grow, if you are not aiming for self-sufficiency I really wouldn't go big on potatoes. They also take a lot of space.

✚ There are few exceptions. New potatoes are such a delight when freshly harvested that I can never resist putting some in.

✚ Blight is the biggest problem in all but the driest climates, so if you do grow your own maincrop potatoes then opt for blight-resistant varieties. At the first sign of blight, remove the foliage (which can go on the compost heap fine). If you have any infected tubers, then these are slightly higher risk. If you have a hot composting system the tubers can be composted but don't leave infected potatoes just on the surface in the garden as they will act as a source of infection for other growers and in following years.

✚ New potatoes can be dug up and eaten immediately; however if you are planning to store your crop you'll need to 'set' the skin first. This means leaving the crop in the ground for a couple of weeks after cutting the tops off. The skins will harden slightly and therefore store better.

✚ To get the maximum crop you will likely need to water your potatoes, but there are a couple of critical periods when watering will really make a difference. The soil should be moist when you plant the tubers to ensure they start growing. You can water them a little during initial growth but giving them too much can encourage pathogens. Once the tubers start forming is another key time. Check this by carefully digging down to a root and looking for the 'hook point'. Once the hook forms, start watering more as the tubers rapidly put on size; without enough water they will not put on bulk, and might be susceptible to diseases such a scab. If possible don't water the leaves as this will encourage blight.

✚ You can successfully grow no-dig potatoes by placing them on the surface of the soil and covering with a thick layer of straw, compost or composted woodchip. So long as the light can't get through and they don't dry out you'll get a great crop with less effort.

ZERO-WASTE TIPS

✚ You can grow potatoes from the peelings. At the end of storage, most potato varieties will be starting to sprout. If you peel those sprouts with a chunk of potato on each one and pot them on, they will start to develop and you then plant them once the last frost has passed.

TOO MANY POTATOES?

✚ Potatoes enter a dormant phase over winter. The best way to store them is to create conditions that make this phase as long as possible. Ideally place them in a cool (5–10°C/41–50°F), dark and damp (95 per cent humidity) area.

✚ Make a clamp for your potatoes (see page 41). They will also store well for a shorter time in paper sacks in a cool damp shed.

RADISH

Raphanus sativus

Radish is one of the easiest crops to grow, though it still needs a little care to get the best crop. As it is very quick to grow, it can be ready to harvest 4–6 weeks after sowing. Radish adds a fresh punch to salads, and pickles beautifully.

Sow: Spring to early autumn, direct. 20 plants per 1m/3ft row with five rows

Plant: n/a

Harvest: 4–6 weeks after sowing, when 2.5cm/1in in diameter

Eat: Fresh, spring to autumn; pickled, winter to spring

Yield

	Per Plant	Per m^2/yd^2
Total	5g/0.2oz	500g/1lb

GROWING TIPS

The great thing about radishes, especially when you are starting out, is that they germinate quickly and well – usually within a week of sowing. You can quickly see whether they have worked. To make sure you make the most of this early promise however here are some tips.

✚ Radishes don't like overcrowding; they won't bulk up if growing too close together. If you end up sowing too tightly, then thin out the seedlings as soon as possible to at least 5cm/2in apart.

✚ Drying out and extreme heat are disasters for radishes. They will quickly get tough and bolt. Except in very cool summers I would grow only in spring and autumn. In the absence of rain then regular watering is essential – give them a good soak a couple of times a week. If grown in pots radishes need water every day in warm weather.

✚ Once ready, radishes need to be picked quickly as they will get tough if left in the ground. This is why sowing small amounts every couple of weeks is the best way to get a regular supply over the whole season.

✚ Also worth noting is that some varieties do better in spring than others – generally you can't go wrong with the small round red ones. Black radishes and daikons are better suited to autumn and winter growing.

ZERO-WASTE TIPS

✚ Radish leaves are edible, though except when very young they have quite a rough texture so are not great in salads. They work well in pesto though.

✚ It is sometimes hard to keep up with the speed of radish growth. If you haven't managed to harvest them all then let them go to seed and harvest the seed pods while still green. They are beautifully spicy and crunchy. They also pickle well particularly in a mix with the root.

TOO MANY RADISHES?

✚ Radish gluts are a distinct possibility since they all tend to mature quite quickly. Pickling is perhaps the best option. You can do a quick pickle with them by slicing the roots thinly, adding some sliced spring onion if you like, and pouring on hot brine. This will last a few weeks till your next radish harvest is ready.

SWEDE

Brassica napus

Sow:	Late spring, direct, final spacing 20cm/8in in rows 40cm/16in apart
Plant:	n/a
Harvest:	Late autumn to winter
Eat:	Fresh, autumn to winter; from store, winter to spring

Swede is one of those vegetables that deserves a comeback. Kale has found a newfound fashion but swede still languishes. Because it's easy to grow, a good keeper and versatile, I recommend growing a few. Grated swede and carrot salad is a favourite of mine.

GROWING TIPS

✚ Swedes take a long time to mature and like most brassicas don't like drying out, so although they are not a difficult crop to grow they do need a little attention.

✚ You can make just one sowing, since swedes mature in late autumn and winter and will sit happily in the ground or in the store till you need them.

✚ Sow quite thickly (every 2.5–5cm/1–2in) as seed is cheap and then thin early to final spacing of 20cm/8in.

✚ A good mulch and watering when very dry will help ensure a good crop.

Yield

	Per Plant	Per m²/yd²
Total	300g/10oz	4.5kg/10lb

PARSNIP

Pastinaca sativa

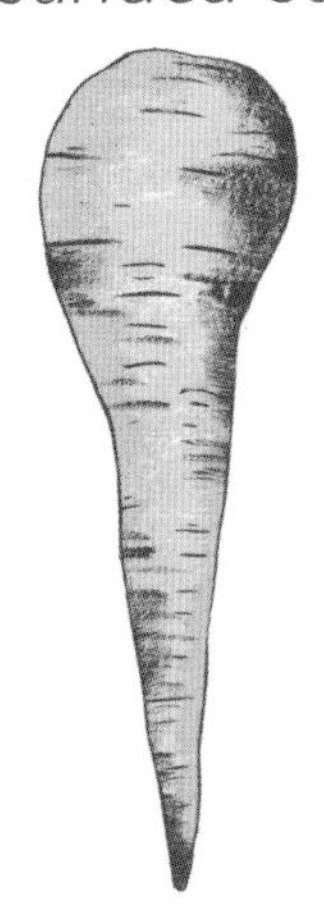

Sow:	Mid-spring direct
Plant:	n/a
Harvest:	Autumn to winter
Eat:	Fresh, autumn to winter; dried/ frozen/stored, winter to spring

Though rich and earthy, parsnips can be really sweet; and Europeans used them as a sweetener before sugar from sugar cane was available. Though germination is sometimes erratic, in good years parsnips grow well and can give a huge yield once they get going.

GROWING TIPS

✚ Seed quality deteriorates with parsnips so use fresh seed, or if using older seed, sow more thickly to compensate.

✚ Despite what many gardening books say, don't sow too early. If the soil is too cold the seed won't germinate but will simply rot or be eaten. For gardening geeks you can go and buy a soil thermometer to check. Or you can do what my old lecturer referred to as the bottom test. If you sat on the soil in your underpants how quickly would you want to stand up again. If it feels cold to you it will be too cold for the plant.

Yield

	Per Plant	Per m^2/yd^2
Total	100g/3½oz	2kg/4½lb

LEEK

Allium porrum

Being tough, hardy and super tasty, leeks are a must in my opinion. With the right varieties, and the weather in your favour, you can eat them fresh from late summer right through to early spring. The beauty of growing your own is that you get the thick green leaf too, which is normally trimmed from the white shank on shop-bought leeks.

Sow: Spring, in a seedbed or modules

Plant: Spring or early summer

Harvest: Late summer to early spring

Eat: Fresh, late summer to early spring; frozen/pickled, spring to early summer

Yield

	Per Plant	Per m^2/yd^2
Total	200g/7oz	3.6kg/8lb

GROWING TIPS

✚ You can sow leeks directly where you want to grow them; however they grow slowly so weeding becomes a full-time job. It is therefore better to grow them on as seedlings and then transplant. They do ok in modules but I prefer to grow the seedlings either in crates of compost, or in the ground. This allows plants to get bigger before transplanting and reduces the pressure to plant out if the weather conditions are not quite right. In modules the plants will get stressed if left too long before planting.

✚ To transplant, use a dibber to make a hole about 15cm/6in deep in the soil, then push the leek seedling into this hole. The young plant will be pretty tough, so don't worry if its roots get damaged as it goes in. Some people even trim the roots and leaf tips before transplanting, to encourage new growth, but that is not essential. Once the plant is in its hole, water it well; this should settle enough soil into the hole to cover the roots and ensure good establishment. The deeper the hole you plant into, the longer the white shank of the leek; you can even earth the plant up as it grows to get more white. Personally I like the flavour of the green too, and am not looking for extra work, so I tend not to bother.

✚ Leeks can get rust (more in dry weather, and watering will help) and mildew (more in wet weather; wider spacing when planting to improve air flow can reduce incidents).

✚ Recently, leek moth has also become more of a problem. The only sure way of dealing with this is a really good crop mesh cover. Strong healthy plants will sometimes grow back after an initial attack.

ZERO-WASTE TIPS

✚ The best zero-waste tip is to eat the whole leaf. It always upsets me how much gets trimmed off commercially. The green leaves are so flavoursome that I much prefer using them in sauces and soups.

TOO MANY LEEKS?

✚ Because most varieties will stand well in the ground until you need them, you probably won't have this issue but if you do . . . soup, soup, soup. I just love leek soup.

✚ You can also pickle leeks, and they freeze fine – though make sure you have cleaned off all the mud before freezing; it's much harder to remove it later.

SPRING ONION

Allium cepa, A. fistulosum

As well as being great for interplanting between slower-growing plants, and providing an early onion flavour as we wait for onions and leeks to be ready during the summer, spring onions are mostly trouble-free and can be harvested small for chive-like leaves.

Sow: Spring to early autumn, direct 2.5cm/1in spacing in rows 20cm/8in apart

Plant: n/a

Harvest: Approximately 8–10 weeks after sowing

Eat: Fresh, late spring to autumn; pickled/frozen, winter to early spring

Yield

	Per Plant	Per m²/yd²
Total	10g/0.3oz	2kg/4.4lb

GROWING TIPS

✚ Along with radishes and lettuce, spring onions are great for beginners and those with limited space. They grow quickly and can be squeezed between other crops.

✚ Spring onions do really well in pots, as they have quite shallow roots and are not heavy feeders. You can even grow them indoors on a sunny windowsill. A pot about 15cm/5in across and 12cm/4in deep would work fine.

✚ I love the onion varieties that form a bit of bulb like 'Musona'. They offer the flexibility of harvesting young as a salad onion, and/or growing on till you get a lovely, mild-flavoured, white bulb. You can sow thickly, thinning out for small ones and leaving the rest to mature.

✚ Downy mildew is a significant problem on spring onions, so in damp climates look out for varieties that claim resistance.

✚ Since they germinate quickly and are not in the ground for long, I tend to sow direct; however spring onions also work well multi-sown in modules and transplanted. If doing by this latter method, then leave a bit more space between clumps when planting, say 10cm/3in.

ZERO-WASTE TIPS

✚ Gathering spring onions little and often is the best way to avoid waste. For many varieties if they aren't harvested promptly the leaves will start to yellow, meaning lots of extra fiddling in the kitchen peeling and preparing, not to mention the wasted leaf.

✚ Spring onions don't last well once picked. Make them last a bit longer by keeping them in a glass of water, like cut flowers; better still if you can store them this way in the fridge.

✚ If you do need to get a load out of the ground at once then it's best to either freeze or pickle them quickly.

TOO MANY SPRING ONIONS?

✚ Spring onions pickle beautifully; they have a slightly milder flavour than most maincrop onions.

✚ You can leave bulbing types to mature a little and pickle as whole baby onions.

✚ Freezing spring onions is quick and easy: just slice them up and pop in a bag or container in the freezer. If you want them to last longer then blanch them first.

GARLIC

Allium sativum

I love growing garlic. It is easy to plant and although you will likely end up with smaller bulb than those sold in the shop, the flavour will be good. Because it grows over winter and is out of the ground by midsummer it can fit in well into the rotation before late-sown crops or green manures.

Sow: Late autumn or winter

Plant: n/a

Harvest: Early summer, fresh (green-topped); midsummer, maincrop

Eat: Fresh, summer; dried, autumn to winter; frozen/pickled, winter to spring

Yield

	Per Plant	Per m²/yd²
Total	50g/2oz	1.4kg/3lb

GROWING TIPS

✚ Garlic needs a two-month cold period after planting to ensure it bulbs up well. For this reason in most climates it's best to plant in very late autumn or early winter. However, you don't want the plant to get too big before the onset of winter or the leaves can get blown about and damaged.

✚ Simply push the individual clove (pointed end up) into the soil till just the very tip is showing. Most books recommend digging a shallow trench but, provided your ground is not compacted, I am not convinced that is necessary. If your soil is that compacted you have bigger problems to worry about than getting your garlic into the ground.

✚ If you haven't managed to prepare your ground in time, don't despair ; you can plant the single cloves in pots, outside. They will root and grow in the pot ready to be planted out any time before spring when planting conditions allow.

Garlic needs a two-month cold period after planting to ensure it bulbs up well. For this reason in most climates it's best to plant in very late autumn or early winter.

✚ Green garlic, in other words before the tops have started to yellow, is one of the first signs of summer in the vegetable harvesting world. You can harvest any time after that, but if you intend to store the bulb then it is best to leave it in the ground until the top has died back.

ZERO-WASTE TIPS

✚ Don't throw out old shrivelled cloves when they sprout,you can use them as normal or you can pop them into a glass of water and grow the sprouts out for use as fresh green leaf for salads or pasta dishes.

TOO MUCH GARLIC?

✚ Can you have too much garlic? I am not so sure! However, if you really need to deal with a glut then of course pickling works well. You can also chop and freeze with oil or butter in ice cubes. This is ideal for just popping a cube into anything that you are cooking.

ONION & SHALLOT

Allium cepa, Allium cepa Aggregatum Group

Good-quality onions are relatively easy and cheap to buy, but I do love having my own. With some careful variety choice and good storage you can grow a supply most of the year, though if you eat as many onions as we do in our house you'd need a biggish garden.

Sow: Late winter or early spring, in modules

Plant: Early spring, plant sets

Harvest: Late summer to autumn

Eat: Fresh, late summer; dried, autumn to early spring; pickled/frozen, spring to early summer

Yield

	Per Plant	Per m^2/yd^2
Total	100g/3½oz	4kg/9lb

GROWING TIPS

✚ Growing onions from sets (specially raised baby onions) is the easiest method. They are straightforward to plant and weed, and should produce a good crop. But beware, they don't store well.

✚ Shallots are grown the same way but rather than forming a big bulb they split into numerous smaller ones.

✚ Growing onions from seed is a little trickier. You need to start the seed earlier than onion sets to give them long enough to bulk up, and because the seedlings are so slight and spindly, keeping on top of the weeding is a challenge. However when successful they are likely to give a better crop and certainly longer storage if you are aiming for total onion self-sufficiency.

✚ Onion seed is best multi-sown in modules. This will give slightly smaller bulbs than if you grow them singly but will make better use of space and is likely to give a greater yield overall. Be careful when transplanting though as they are fragile.

✚ If you want the earliest cropping varieties then go for overwintering Japanese onions. They are usually ready a couple of weeks before 'normal' types. I gave up on them quite early on as you have to look after them all winter, protecting them from the wind, the pests and the snow.

✚ Red onions tend to have a milder flavour than white, though this will also depend on the season. A hotter drier year can produce stronger-flavoured onions. If you prefer them sweeter then keep them watered.

✚ Onions need both long days and warm temperatures to reach their full potential, though they prefer it to be cooler earlier in the season when they are putting on their initial growth.

ZERO-WASTE TIPS

✚ Green onions are a useful way to start eating your crop as soon as possible. If you wish to store your onions it's best to wait till the tops have died back, but that doesn't mean you can't start pulling them up before then. The tops are delicious and can be finely sliced as a chive alternative in salads or as pizza topping.

✚ If some of your onions bolt, don't give up on them. You can still eat them but you need to get them out quickly before they take all the energy from the bulb and use it to make their flower and seeds. If you don't get there soon enough there's still a way to eat them, and that is to harvest the buds or very young flowers. These can be braised or pickled.

TOO MANY ONIONS AND SHALLOTS?

✚ Apart from the few we eat fresh, most onions are stored. The key is to keep them warm and dry. Their dry skin is perfect natural protection, and unlike root crops or apples we need to keep the air moving around them while stored. Be aware though that you won't see if they are rotting and need to feel them regularly to check.

✚ Pickled onions are delicious and a great addition to mixed pickles, but onions dry so well that you don't need to pickle or freeze them unless you have a load about to go off or sprout. This is more likely if it has been a wet season or if they have suffered from fungal infections.

AUBERGINE/EGGPLANT

Solanum melongena

Aubergines are a little temperamental to grow, but very satisfying when they work well. They normally need protection in cooler climates, though new varieties and our warming climate mean that outdoor growing is becoming a possibility. I particularly like the baby and thin varieties as they need a shorter ripening period.

Sow: Late winter or very early spring, heated

Plant: Late spring or early summer

Harvest: Summer

Eat: Fresh; summer, pickled/chutney/frozen, year-round

Yield

	Per Plant	Per m²/yd²
Total	900g/2lb	2.75kg/6lb
Per pick	300g/10oz	900g/2lb

GROWING TIPS

✚ Aubergines need a long season and plenty of warmth. This makes them trickier to get right in cool climates but not impossible, provided you have heated propagation space and greenhouse or polytunnel space.

✚ The ideal germination temperature for aubergines is 25–30°C/75–85°F. You could still get some germination when it is as low as 16°C/61°F, but it will be slower and the percentage of seeds germinating may be lower.

✚ Start aubergines as early as you can – in late winter or very early spring – to give them a good start by the time you plant them out.

✚ The plants can grow big and though they don't climb they will need some support as the stems are prone to breaking once the weight of the fruit is on them. Some varieties will cope with just a stake holding the main stem; others might need more. The plant supports used for some ornamental perennial plants work well, or you can make your own with some wire mesh held horizontally about 45cm/18in off the ground.

Start aubergines as early as you can – in late winter or very early spring – to give them a good start by the time you plant them out.

✚ Plants need 60cm/24in between them. Keep them well watered. It's even worth spraying the leaves (or the paths of the greenhouse) to keep the humidity levels up.

✚ It is usually recommended to limit the number of fruits you allow to mature on each plant. Though this will depend a little on your soil health and spacing, there is a chance that if you let too much fruit develop you'll end up with small fruit, or that some of them won't form at all. For larger-fruiting varieties, five aubergines is a good number to aim for. Often the plant will self-regulate and the flowers will drop off if it doesn't have the energy to produce more fruit.

TOO MANY AUBERGINES?

✚ Aubergine makes a lovely spiced chutney, but also pickles well. You'll need to salt the aubergine to draw the moisture from it, and cook it for a couple of minutes before pickling.

✚ This vegetable works well frozen, but is best roasted or grilled before freezing. Slice into thin discs or strips (depending on the varieties) and grill until soft. Freeze in small portions that can be added to sauces or casseroles from frozen.

BROAD (FAVA) BEAN

Vicia faba

The hardiest and earliest of the beans to grow are broad beans, which are very easy to germinate and in most years reliable croppers. New shorter varieties don't even need staking. They can tolerate light shade, especially early-sowing ones that do most of their growing before many trees are in full leaf.

Sow:	Late winter or very early spring, heated
Plant:	Late spring or early summer
Harvest:	Summer
Eat:	Fresh; summer, pickled/chutney/frozen, year-round

Yield

	Per Plant	Per m²/yd²
Total	100g/3½oz (pods)	1.6kg/3½lb
Per pick	25g/1oz	400g/14oz

How often to pick

Pick every three to five days.
This interval will be longer in dry conditions.

GROWING TIPS

✚ Broad beans have big seeds and germinate quickly (often in less than a week). They will also germinate in low temperatures, even as low as 7°C/45°F though they prefer it a little warmer. The plants themselves are also fairly hardy.

✚ You can sow them either in late autumn for an early crop the following year or start them in early spring for a slightly later harvest. As with overwintering onions I mostly don't bother with the hassle of looking after them over winter and just go for a spring planting.

✚ Sow double rows of seed leaving 15–20cm/6–8in between seeds and 60cm/24in between double rows.

✚ Older varieties tend to be taller, though sometimes more productive per plant, and will need some support. I put stakes around the whole patch or line of plants, and tie a couple of lines of string at around knee and waist height, which stops too much wind blow.

✚ You can put an individual cane for each plant but I find they don't need that much support.

✚ Broad beans like it cool and need plenty of water. In hot springs and summer you will definitely need to water them once they start producing pods to ensure you get a decent crop.

ZERO-WASTE TIPS

✚ Broad bean leaf shoots are delicious, and removing them before blackfly appear is said to help reduce blackfly numbers, which are one of the main pests on broad beans. Pinch out the tops when plants are about 60cm/24in high. This will also encourage a bushier stronger plant.

TOO MANY BROAD BEANS?

✚ Broad beans are also called fava beans and can be used in pretty much any way you use chick peas. Hummus, soups, falafel and more can all be made from the delicious fresh or dried bean.

✚ Drying is the simplest way to store broad beans for a long time. Dry them in the pod, laid out single depth on a tray or rack in a warm dry place such as a greenhouse, polytunnel or on a sunny windowsill. Once the pods are dry and brittle you can rub them together to get the dried seed out. I usually then dry the beans on their own for a bit longer before storing.

GREEN BEANS

Phaseolus vulgaris	Climbing green beans
Phaseolus vulgaris	Dwarf green beans
Phaseolus coccineus	Runner beans

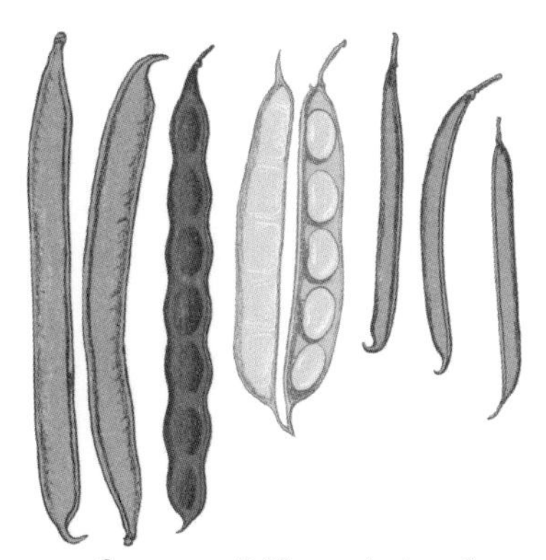

From super-fine, pencil-like varieties, through flat beans to the larger runners that can grow up to 50cm/20in long, the range of shapes and colour of green bean is wonderful. This group also includes most of the drying beans. I certainly wouldn't be without them in my garden.

Sow:	Late spring, direct or in pots/modules
Plant:	Early summer after last frost
Harvest:	Late summer to early autumn
Eat:	Fresh, summer to early autumn, frozen/pickled/dried, winter to spring

Yield

	Per Plant	Per m²yd²
Total	150–200g/5–7oz	3.6–4.5kg/8–10lb
Per pick	25g/1oz	600–800g/1¼–1¾lb

GROWING TIPS

✚ Broadly all these types of beans need the same thing in terms of soil, climate and water. Being members of the legume family, they fix their own nitrogen so don't need masses of fertility. They like to be in well-drained soil, but also require lots of water once the beans start to form.

✚ Once the bean pods start to form, keep picking regularly. If you leave beans for too long the plants will think their seed-producing work is done and put the effort into growing and ripening rather than producing more flowers. The exception to this are borlotti beans: just let them grow without picking until all the beans are fat and the pods start to look dry.

✚ Though there is some variation between varieties (for example, runner beans need a bit more room), space at about 15cm/6in between plants with 30–40cm/12–16in between rows.

✚ The other big difference between varieties is whether they form a bush or are climbers. Flat and pencil beans can be either bush or climbing, but runner beans are all climbers. Bush varieties will not need any support. Climbing varieties of course need something to climb – poles, trellis and maize plants are all possible.

✚ Though most people put their pole supports in an A shape, the beans are actually easier to pick if you create an X shape. The fruit after all hangs down from the plant and in an X shape they are hanging outside the structure where you can see and reach them rather than in the middle, where they tend to be hidden.

✚ Climbing varieties usually produce more beans than bush ones but over a longer period. For early sowings of bush varieties you can wait until most beans are a decent size and then pull the whole plant up, harvest the beans and get another crop of something in afterwards.

✚ Although some varieties like borlotti have been specially bred for drying, you can actually save and dry any bean. I often stop picking some of the plants towards the end of the season and let the plants produce a final crop for drying. This also works well if you are planning to save your own seed for growing next year's crop.

If you leave beans for too long the plants will think their seed-producing work is done and put the effort into growing and ripening rather than producing more flowers.

How often to pick

Pick every two to four days.
This interval will be longer in dry conditions.

ZERO-WASTE TIPS

✚ Don't use plastic or nylon string for supporting or tying in your poles. Apart from wanting to reduce plastic use, if you use a natural compostable string such as hemp then at the end of the season you can just pull off the entire plant and string in one go and put it all in the compost.

✚ Pick beans when they are young. Although you might end up with slightly lower yield, you are less likely to have gluts that get wasted, and more importantly you don't have to waste effort, and crop, by removing the stringy edges of the beans (this is particularly a problem on runner beans and some flat varieties).

✚ If you do find at the end of the season that you've not kept up with the picking then leave the beans on the plant to use for drying.

TOO MANY GREEN BEANS?

✚ I have been known to eat a plate mounded high with beans, just smothered in butter with salt and pepper. However plants can produce a lot in a good year so sometimes even mass consumption doesn't use them all.

✚ Leaving some plants unpicked for dried bean production is the easiest way to extend the harvest.

✚ Beans pickle beautifully, either whole if relatively small, or sliced if larger. They also freeze quite well, though I always think they lose something of their delicious crisp texture after freezing; they seem to go just a little soft.

JERUSALEM ARTICHOKE

Helianthus tuberosus

Sow:	Winter as tubers
Plant:	n/a
Harvest:	Winter
Eat:	Fresh, winter

This is probably the most reliably productive vegetable you can grow. It suffers from almost no pests and diseases and is very productive. The biggest problem you are likely to have is how to control it.

GROWING TIPS

✚ Choose a smooth-skinned 'Fuseau' variety. The knobbly ones are so fiddly to prepare you will get fed up with eating them.

✚ If left in the same place every year Jerusalem artichokes will grow fine but the tubers will tend to get gradually smaller. Therefore transplant a few to a new place every two or three years. Make sure you dig out all the tubers when you move them or they will keep coming back.

ZERO-WASTE TIPS

✚ Jerusalem artichokes are related to sunflowers and produce a yellow flower that works well as a cut flower.

TOO MANY JERUSALEM ARTICHOKES?

✚ Soup is my favourite way of eat them and uses a lot.

Yield

	Per Plant	Per m²/yd²
Total	2kg/4½lb	12kg/26lb

COURGETTE/ZUCCHINI

Cucurbita pepo

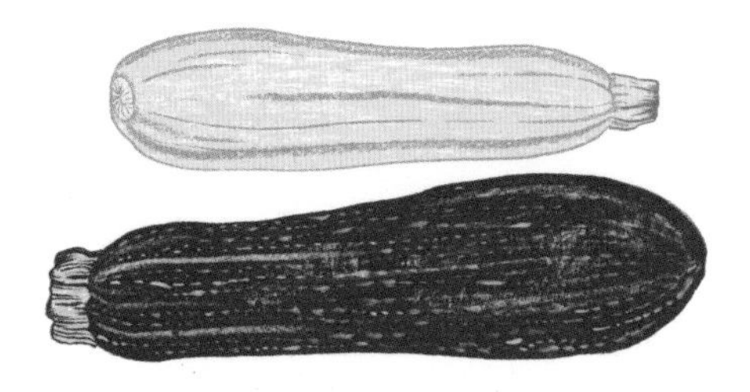

Forget about accurate yield prediction when it comes to courgettes. At the start of the season I am desperately willing my plants to produce their first fruit, only to be screaming 'no more please' as they get larger and more numerous. Finally as they succumb to inevitable mildew and death at the end of the season I find myself mourning their passing.

Sow:	Spring, in pots; early summer, direct
Plant:	Late spring to early summer
Harvest:	Summer to autumn
Eat:	Fresh, summer to autumn; pickled/frozen, winter to spring

Yield

	Per Plant	Per m^2/yd^2
Total	3kg/6½lb	9kg/20lb
Per pick	200g/7oz	600g/1¼lb

How often to pick

Pick every one to four days.
In the right conditions they grow very quickly

GROWING TIPS

✚ Direct-sown plants will often do well, and even outgrow transplanted ones, though they depend on a warm spell and no slugs. For this reason they tend to be started in pots inside to give them a head start. Courgettes are especially susceptible to wind for the first couple of weeks after planting out, since the seedlings have got used to the easy life in the greenhouse. It is very important to harden them off well. One other trick is to brush the seedlings gently with your hand as they are developing. This mimics wind and helps strengthen the plants as they grow.

✚ Courgette plants can grow big, so give each one at least $1m^2/yd^2$ to grow.

✚ Courgette fruit grows very quickly, and while marrow has its fans I prefer mine small. Pick courgettes regularly to avoid being overwhelmed by huge piles of marrows.

✚ In very dry conditions courgette plants need watering regularly, or the fruit will not properly develop.

✚ Courgettes are suitable for undersowing with a low-growing, white clover or yellow trefoil. Just keep an eye out in the early stages to make sure the green manure isn't encroaching too close to the plant. Once the courgette gets going it will easily outcompete.

✚ Yellow courgette varieties are slightly less vigorous than green ones and tend to produce smaller fruit.

ZERO-WASTE TIPS

✚ You can eat courgette flowers. Stuffed with creamy fillings or deep-fried in batter are just two of the high-calorie ways to enjoy them. You can add them to pasta or stir-fry dishes too. There is a trade-off when picking the female flowers since you have to pick the courgette very small or the flower will have rotted off. The male flowers, however, are a bonus since they would not otherwise be used.

✚ Courgettes and squashes grow really well on the compost heap, assuming that you have a pile of 'maturing compost' rather than the one you are actively adding to.

TOO MANY COURGETTES?

✚ Courgettes don't have a strong flavour and are mostly water, so cook down to not very much. You can add them instead of liquid to many sauces, stews and soups.

✚ When grated courgettes can also be added to cakes or salads. The fashion of spiralizing is also very well suited to courgettes.

✚ Courgettes will freeze but tend to come out pretty mushy at the other end. This is fine for soups and casseroles, but not very exciting as a vegetable on their own.

CUCUMBER

Cucumis sativus

Though available in supermarkets year-round, there is really nothing like a freshly picked cucumber, or a mound of tzatziki made with your home-grown dill. There are great outdoor varieties but the most productive ones usually need protection.

Sow:	Spring, in pots
Plant:	Late spring or early summer
Harvest:	Summer to early autumn
Eat:	Fresh, summer to early autumn; pickled/frozen, year-round

Yield

	Per Plant	Per m²/yd²
Total	3kg/6½lb	12kg/26lb
Per pick	300g/10oz	1.25kg/2¾lb

How often to pick

Pick every two to four days.

GROWING TIPS

✚ Cucumber plants like high humidity. This makes it hard to give them the perfect conditions in a greenhouse alongside tomatoes, which need the opposite. Spraying the cucumber leaves regularly with water will help, and in particular will reduce the risk from red spider mites, which love both cucumbers and the hot dry conditions of a greenhouse.

✚ Though you may be desperate to get some fruit from your plants, it helps longer-term plant health to remove the first few flowers until the plant is 60cm/2ft tall, as it allows it to put all its energy into growth.

✚ There are lots of theories on pruning cucumbers, and even commercial systems differ. Broadly speaking, the less you thin and prune your plant the greater the potential yield, but the higher the risk of disease. Pruning more side shoots removes leaves, which means the plant can support fewer fruit, but also increases air flow, which reduces the potential for fungal infection. If you want to prune hard, then pinch out every side shoot as soon as it has one or two fruit on it. Leave a couple of leaves beyond each fruit.

ZERO-WASTE TIPS

✚ You can eat cucumber flowers fresh in salads or they look nice as a summer drinks garnish.

✚ If you are thinning out the fruit to prevent overcrowding on the bush then don't throw away the baby fruitlets – they are great as a crunchy snack.

TOO MANY CUCUMBERS?

✚ You can freeze cucumbers, but they come out squishy and not suitable for salads. You can still use them for juicing, tzatziki or gazpacho.

✚ Pickled cucumbers are great. Either choose specific pickling baby cucumbers or slice up the bigger ones.

PEA

Pisum sativum

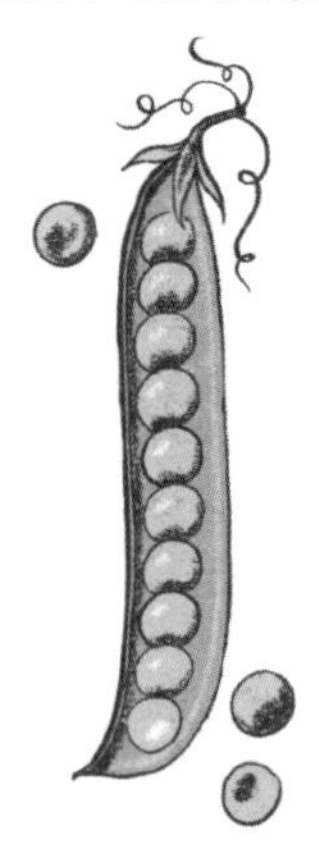

Peas quickly lose their sweetness when picked so I normally buy frozen peas, which are processed and frozen in the field within a couple of hours of harvest. However eating your own fresh from the bush is another matter altogether. Most of our peas don't make it as far as the kitchen.

Sow:	Early spring to late summer, direct or in pots/modules
Plant:	Late spring to late summer
Harvest:	Summer to autumn
Eat:	Fresh, summer to autumn; dried/frozen, winter to spring

Yield

	Per Plant	Per m²/yd²
Total	100g/3½oz	2kg/4½lb
Per pick	20g/¾oz	400g/14oz

How often to pick

Pick every two to four days.
This interval will be longer in dry conditions.

GROWING TIPS

Peas are really easy to grow, but slow and fiddly to pick and prepare.

✚ Sow thickly 5–10cm/2–4in apart in double rows 45cm/18in apart. Though you may not need that many plants, I find that I usually lose some plants and once they start climbing they find their own space. Observe how many grow for you and increase spacing the following year if needed.

✚ Mice can be a problem. They really like eating pea seeds both in the ground and even in pots. If you have a major problem with them you will need to cover your module trays with mesh or put them on racking that mice can't climb.

✚ Peas, like green beans, need regular picking to keep them producing and they will quickly get tough if left to get big. Make sure you get all the pods off – yes even the ones hiding down at the bottom in the middle of the rows – otherwise the plant will enter ripening phase and you'll get a reduced harvest.

✚ Harvest mangetout or sugar snap varieties young, before they get tough.

ZERO-WASTE TIPS

✚ Pea shoots – the bit that will eventually become the flower – are delicious in salads. I used to make good money from them when growing for chefs. Even better you can get a good crop of them when the plants are young without seriously affecting your pea crop – in some cases it even helps the plant bush out a bit. I usually grow a succession of peas and pick shoots from each sowing for a couple of weeks before allowing it to produce peas.

TOO MANY PEAS?

✚ In our house leaving a pile of fresh pods on the kitchen table is the surest way to clear a glut – the family all pick at them as they go by.

✚ However, if you have a massive pile, then peas pickle perfectly; mangetout and sugar snaps work especially well; and they all freeze well too. To ensure maximum flavour, just be sure to process peas as quickly as possible after harvest.

PEPPER

Capsicum annuum

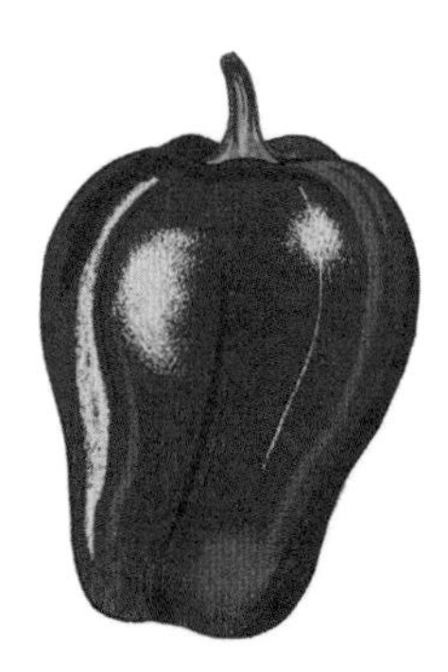

There's a huge range of capsicum, or peppers, from the almost uneatable super-hot chillies to the sweet mild bell peppers that dominate supermarket sales. The joy of growing your own is the chance to try types you never find in the shops.

Sow:	Early spring, heated
Plant:	Late spring, under cover; summer, outdoors
Harvest:	Fresh, summer to autumn
Eat:	Fresh, summer to autumn; pickled/dried/frozen, year-round

Yield

	Per Plant	Per m²/yd²
Total	1.5kg/3¼lb	4.5kg/10lb
Per pick	250g/9oz	750g/1½lb

GROWING TIPS

✚ Peppers like it warm, right from the off. Give them 18–24°C/65–75°F for germination, and keep them warm especially during the cold nights.

✚ Peppers operate on an average temperature between night and day, so even when the days get hot if the nights are cold they will grow slowly. Keeping the plants warm at night really helps keep them growing.

✚ Suggested yields here are based on large sweet peppers; for chilli peppers there is so much variation that it is hard to generalize. For the small hot varieties you may need only one bush to give you all the chillies you need. For milder ones I normally grow two or three plants. 'Hungarian Hot Wax' is one of my favourite moderately hot chillies to grow as it copes well with cooler climates and is really prolific.

✚ Peppers grow well in a large pot, though you may need to add some liquid feed as the season goes on, and keep them well watered. The advantage of pot-grown plants is that you can move them into the outdoor sunshine early in the season, and bring them back indoors at night to prevent chilling.

ZERO-WASTE TIPS

✚ The seeds of peppers are edible but are extremely hot and can be bitter.

✚ There is anecdotal evidence that spreading ground-up chilli around plants will keep off some pests like mice. Use your old or damaged fruit to keep the unwanted visitors away.

TOO MANY PEPPERS?

✚ Gluts of sweet peppers freeze well: either just cut up and freeze on trays (blanching for longer shelf life) or roast them first. The advantage of roasting first is not only that extra flavour, but that they will also take up less room in the freezer.

✚ Sweet and chilli peppers are great pickled or fermented, and can be eaten either as they are or as an ingredient in other dishes. It's best to label which varieties you used to avoid adding too much of a super-hot variety – I say this from painful experience.

✚ The easiest way to keep chilli peppers is to dry them. You can hang the whole plant upside down in a warm dry place and then just pick dried fruit as you need them. Once they are properly dry, store them in an airtight drawer.

PUMPKIN

The golden rule for a zero-waste pumpkin is to make sure you grow a tasty variety. Many of those grown for lanterns have a watery disappointing flavour. Choose for taste and you can eat well and still make your lantern for Halloween.

Sow:	Spring, indoors; very early summer, outdoors
Plant:	Early summer
Harvest:	Autumn
Eat:	Stored, autumn for 2–5 months

Yield

	Total pumpkins per plant	Weight per pumpkin	Average weight per plant	Per m²/yd²
Mini	5–10	500g/1lb	3kg/6½lb	4
Medium	2–5	4kg/10lb	12kg/26lb	7
Large	1–2	8kg/17½lb	14kg/31lb	9

GROWING TIPS

✚ Pumpkins need a surprising amount of space. The largest varieties require 2.75 × 2.75m/9 × 9ft for each plant. Mini varieties such as 'Jack be Little' can manage with 1.75 × 1.75m/6 × 6ft. There are ways to pack them in if you don't have a massive garden: for example, grow quick crops like spring onion or salad around the plants when they are young and harvest them as the pumpkins grow to fill the space; grow the pumpkins among taller plants like runner beans and sweetcorn; grow mini varieties up trellis for a vertical crop – the larger types are too heavy for this method and even the small ones may need supporting; or grow a cover crop like white clover or yellow trefoil under the pumpkin, that way once the plant dies off in the first frosts you still have green cover on your nitrogen-fixing soil, capturing sunlight and holding on to nutrients.

✚ Pumpkin plants often do better if sown directly so long as the soil is warm enough (18°C/65°F) early enough to give the fruit at least 100 days before the first frosts come in autumn. For those with shorter colder summers you can sow into pots indoors in the heat and then plant out after the last frost. Make sure you protect young plants from the wind as they are susceptible for the first couple of weeks.

✚ Pumpkins need a lot of water, especially if you want big fruit. Water at planting to establish, and regularly once the fruit has formed.

✚ Most pumpkin varieties need at least 100 days from sowing before you can harvest them. Some need as many as 120 frost-free days to grow. Be patient; if you harvest too soon the fruit will not store well.

ZERO-WASTE TIPS

✚ You can eat pumpkin seeds, but there is a catch. Some varieties have been specially bred to have seeds with a soft coat that don't need shelling. The flesh from some of these types is not very good to eat – it is usually pretty tasteless and stringy. The seeds from most eating pumpkins have a shell that is edible but pretty crunchy. You can roast them with oil, salt and spices, and they're pretty good, but you need a good set of teeth. Look out for varieties like 'Streaker', Hull-less or 'Eat All' to get both 'naked' seed and tasty flesh.

✚ Eat pumpkin flowers raw in salads, or stuffed and baked, or even deep fried. Pick male or female flowers, though make sure you have let some fruit set first. Some varieties can be bitter, so check one flower before serving them all up to your friends.

TOO MANY PUMPKINS?

✚ Pumpkins store well whole for months in the right conditions. If you still haven't managed to eat them all, then there are lots of ways of preserving them.

✚ Pureed pumpkin freezes well, and you can dry cubes of the fruit to use in soups through the year.

✚ You can even make pumpkin wine.

TOMATO

Lycopersicon esculentum

Tomatoes are one of the tastiest and most versatile crops you can grow. Mix up your varieties, and include both cherry and beefsteak types, to get the maximum benefit from all their different uses, from salad to sauce.

Sow:	Early spring, indoors
Plant:	Late spring, indoors; summer, outdoors
Harvest:	Summer to autumn
Eat:	Fresh, summer to autumn; frozen or processed, year-round

Yield

	Per plant cherry	Per plant round/beefsteak	Per m^2/yd^2 cherry	Per m^2/yd^2 round/beefsteak
Total	4kg/9lb	5kg/11lb	16kg/35lb	20kg/44lb
Per pick	200g/7oz	250g/9oz	800g/2lb	1kg/2¼lb

GROWING TIPS

✚ Sow bush varieties to avoid having to pinch out the side shoots. They're also good for outdoor growing as they tend to fruit a bit quicker.

✚ If you want to grow tomatoes outdoors look out for a short 'days till fruiting' time (often given on the packet). The quicker the plants come to fruit the more likely they are to ripen outside in cooler climates.

✚ Use red plastic mulch – the light it reflects helps to increase sugar levels in the plants, which improves the flavour.

✚ Pick approximately every two days, though you can pick less often if you pick some that aren't totally ripe and finish ripening them indoors.

ZERO-WASTE TIPS

Everyone has some tomatoes that don't ripen, especially if you live in a cooler climate or one with a short summer. Here are some ideas for using green tomatoes at the end of the season (or at any time):

✚ **Green tomato relish** – you can substitute green for red in most chutney recipes but you might lose some of the sweetness and richness. Mixing half and half is a good way of using the green without compromising too much on taste.

✚ **Fried green tomatoes** – slice thinly, cover in batter and deep- or shallow-fry. You might find yourself growing green tomatoes just so you can eat more of them this way.

✚ **Slow-baked green tomatoes** – this works especially well if you bake the tomatoes alongside a joint of meat, but cooked in olive oil is also delicious. Larger fruit should be quartered or halved.

TOO MANY TOMATOES?

If you get good at growing these and have a few plants, there may well be times of the year when you have too many tomatoes to eat in your salads or on your pizzas. Preserving tomatoes to keep you going through the winter months is the best way to make use of them. Here are just a few ideas:

✚ **Bottled tomatoes** – because tomatoes are highly acidic, they keep well without much preparation. Add a bit of salt and lemon to the bottom of the jar, almost fill the jar with tomatoes, pour boiling water over them and pop the lid on. Then put the bottle in a hot water bath for about 45 minutes.

✚ **Passata** is so easy to make and much better than the bought stuff. It's just cooked tomato sauce, put through a sieve or blender.

✚ There are many ways to freeze tomatoes but the following is probably the easiest. Lay cherry tomatoes out on a tray and stick the tray in the freezer. The tomatoes will freeze whole and you can then transfer them to a bag or tub. When you need to use them, you can just pick out as many as you need one by one. For bigger fruit, cut into pieces and freeze, or cook them into a tomato sauce and freeze in useable portions.

SWEETCORN

Zea mays

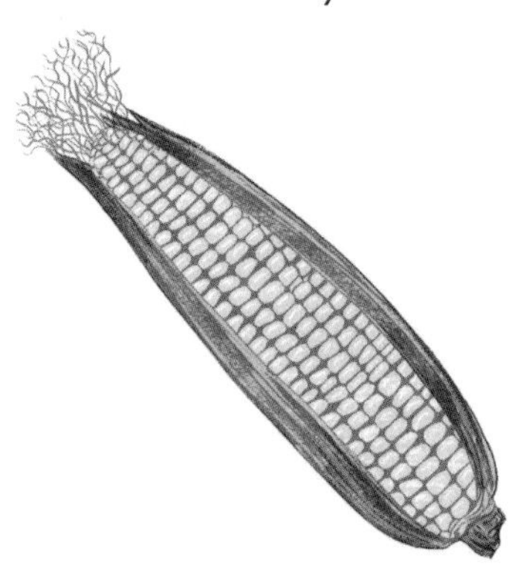

New varieties with shorter maturing mean that growing delicious sweetcorn even in cooler climates is now possible. They need sun and good soil but otherwise are relatively easy to grow. Eat quickly after harvest to maintain sweetness.

Sow: Spring, in pots indoors; early summer, outdoors direct

Plant: Early summer

Harvest: Late summer to autumn

Eat: Fresh, late summer to autumn; frozen/pickled, year-round

Yield

	Per Plant	Per m²/yd²
Total	1 cob	9 cobs
Per pick	1 cob	3 cobs

GROWING TIPS

✚ Sweetcorn does best if sown directly in the soil; however you do need to have warm springs to do this. It likes soil temperatures of 15°C/59°F to germinate well. In cooler climates starting under cover in pots or large module trays is the only way to ensure a good crop for some varieties.

✚ Plants take 60–100 days to mature. Mostly this is dependent on variety, and the packet should state how long that variety needs. Weather will also play a part of course, and particularly in the ripening stages sweetcorn definitely needs plenty of sunshine.

✚ Though I have said you will get one cob per plant, this is actually a bit pessimistic and you might well average one and a half cobs per plant, with the second ripening slightly later than the first.

✚ Corn is wind-pollinated, which means the pollen is blown from male flower at the top of each plant to the female cob tassels on other plants, relying on chance to land in the right place. Each strand of the tassel will form one kernel of corn. If you end up with cobs where only half the kernels have formed it's because pollination was poor. To get good pollination you'll need a few plants grown in a block, though if there is lots of sweetcorn growing around you this is less of a problem.

✚ Harvest when the tassels have dried to a brown colour.

✚ Like peas, the sugar in sweetcorn starts to break down quickly, so the sooner you can get the cobs from the plant to the pan, barbecue or freezer the better.

ZERO-WASTE TIPS

✚ A traditional South American method of space-saving is to grow climbing beans up the sweetcorn (with trailing squashes covering the ground). This 'three sisters' system works well in dry climates but not so well in wetter conditions when everything grows so vigorously that the leaves can compete with each other.

TOO MUCH SWEETCORN?

✚ Sweetcorn tends to come all at once. All preservation methods require first stripping the corn from the cob – easy to do with a sharp knife, though you can buy special tools to do the job. It freezes well but is also superb in chutneys and relish.

✚ You can pickle/ferment sweetcorn. As it is naturally high in sugar it gives one of the sweeter pickles. It is great with some chilli to add a spicy bite to the sweetness.

BASIL

Ocimum basilicum

Nothing says summer like the smell of freshly cut basil. I love it sprinkled fresh on to a pizza as it comes out of the oven, but it works well in most salads and pasta dishes too. Basil likes warm growing conditions but is otherwise very easy to grow.

Sow: Spring to early summer, indoors, in modules or pots

Plant: Early spring, under cover; late spring or summer, outdoors

Harvest: Early summer, indoors; late summer to early autumn, indoors

Eat: Fresh, summer to early autumn; frozen/bottled, year-round

Yield

	Per Plant	Per m²/yd²
Total	50g/2oz	1.25kg/2¾lb
Per pick	150g/5oz	3.6kg/8lb

How often to pick

Pick every one to threes weeks.
This will be longer when weather cools.

GROWING TIPS

✚ Basil really does not like the cold: it needs at least 20°C/68°F at germination right through initial growth to maturity. Try to temper your enthusiasm for fresh basil and wait for the conditions to be right before planting out. I have had more than one early crop of basil either fail or take ages to grow, because I was too impatient to get them in the ground.

✚ For indoor crops, leave 20cm/8in between plants. Outdoor crops tend to grow a bit smaller so you can plant slightly closer, at twenty-five plants per m^2/yd^2.

✚ When grown in soil, and if the weather is not too hot, basil will last well and can be picked every week or two over a six-week to two-month period. You should get more than three picks if you can prevent it flowering. If you just take a few leaves off each pick, you can harvest more frequently. Cutting longer stems means you need to leave more time for it to regrow. Picking more often will also encourage regrowth but overpicking will send the plant into shock and can make it bolt.

✚ In very hot weather, water basil plants every two or three days to keep them from bolting – however they do like it hot and relatively dry so don't overwater.

ZERO-WASTE TIPS

✚ When basil bolts, and it will at some point, that isn't the end of the crop. You can still eat the flowers. They are a bit milder than the leaf but still lovely. You can include the tips of the stem but lower thicker parts will be tough.

✚ Have you ever bought one of those super-looking basil plants from the shops only to be disappointed when they don't last? This is because they sow lots of seed into each pot to make it look full, but there are not enough nutrients in the small pot to keep them going for long. The trick is to immediately divide them up into three or four and repot each section into separate larger pots or into the ground. That way you'll be able to keep picking for much longer.

TOO MUCH BASIL?

✚ My go-to-basil glut recipe is of course pesto, which you can freeze quite happily. However if you don't have all the ingredients handy for that, just whoosh up the leaves with lots of oil or butter and freeze.

MEDITERRANEAN HERBS

Salvia Rosmarinus Rosemary
Thymus Vulgaris Thyme
Origanum Vulgare Oregano
Salvia officinalis Sage

These pungent hardy herbs are the backbone of Mediterranean cooking and great, easy plants to grow. Rosemary, sage and thyme keep their leaves throughout winter, giving you fresh flavours year-round.

Sow:	Spring; or take cuttings in late spring
Plant:	Autumn
Harvest:	Fresh, year-round
Eat:	Fresh, year-round; dry or freeze with oil

Yield

	Per plant Thyme	Per plant Sage/ Rosemary/Oregano	Per m^2/yd^2 Thyme	Per m^2/yd^2 Sage/ Rosemary/Oregano
Total	40g/1½oz	200g/7oz	320g/11.25oz	800g/1¾lb
Per pick	2g/0.07oz	10g/½oz	16g/0.5oz	40g/1½oz

How often to pick

Pick as needed, but regular picking ensures new young tender growth'

GROWING TIPS

✚ Rosemary is tricky from seed, but the other three herbs described here work well, given some heat. However since you don't need lots of plants it is often easier to leave the propagation to professionals and buy just a couple of each herb.

✚ Once you have them it is easy to take semi-ripe cuttings in the spring and get new plants that way. Oregano being a herbaceous perennial can be divided in autumn. Being short-lived perennials, you will need to replace every few years, or the plants get woody at the base and less productive.

✚ To ensure a fresh supply of younger leaves you'll need to keep cutting even when you don't need the herb. This will also prolong the life of the plant and keep it from getting straggly. This is especially true of thyme, which can quickly get woody, making it fiddly to use the herb. One easy way of doing this is to have three plants of each herb and to trim one back every few weeks, so you have a bush at different stages of growth.

✚ These plants are used to growing in poor soil and don't need watering in most climates. However, they do need sun.

✚ Look out for variations such as lemon thyme and pineapple sage for some different flavours in your dinner.

ZERO-WASTE TIPS

✚ Don't forget to eat the flowers! The flowers from all these herbs are edible either raw or when added as flavouring. This means you can keep picking even when the plants are blooming. I particularly love thyme flowers – they have an additional peppery note on top of the thyme flavour.

TOO MANY HERBS?

✚ In order to make sure you always have these herbs when you need them, you are likely to have far more than you need. You can of course add them to most pickles, stews and soups. However I tend to treat them as ornamental and insect attractants and not worry about using them all.

✚ Fried sage leaves are delicious. Fry very quickly (10–20 seconds) in hot shallow oil. You can eat them as they are, or sprinkle on to salads or soups.

✚ Freeze the flowers in ice cubes for a decorative addition to your summer drinks.

PARSLEY & CORIANDER

Petroselinum crispum, Coriandrum sativum

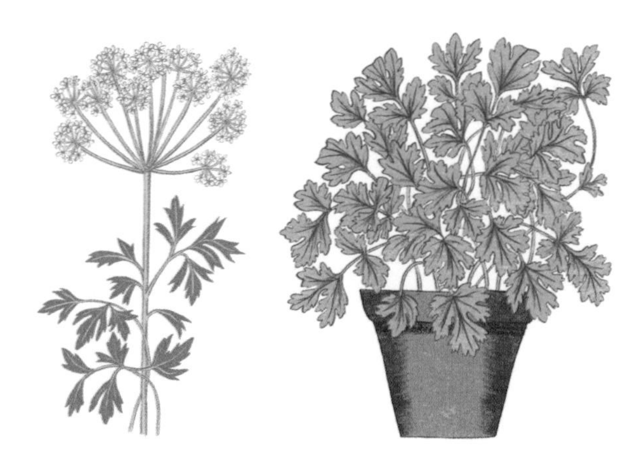

Though not exactly the same of course, coriander and parsley need broadly the same treatment when growing for the leaf herb. Successional sowings will give you a continuous supply of fresh leaf. In addition you can grow some parsleys on for a root crop, and you can grow both for seed.

Sow:	Early spring to early summer, direct or in modules; or autumn
Plant:	Spring or summer; or autumn
Harvest:	Spring to autumn
Eat:	Fresh, spring to autumn; frozen or bottled, year-round

Yield

	Per Plant	Per m^2/yd^2
Total	150g/5oz	3.6kg/8lb
Per pick	50g/2oz	1.25kg/2¾lb

GROWING TIPS

✚ Sow often and water well is the best tip for these herbs.

✚ Coriander is an annual and is particularly prone to bolting when dry and/or hot. Look out for varieties that claim to be slow bolting – though they might not always deliver. Plants need to be spaced at about 20cm/8in in rows 20–30cm/8–12in apart. Coriander is a great plant for interplanting with slower-maturing vegetables as it germinates so quickly. The seed is generally cheap so if the plants are getting in the way of the main crop, I don't mind sacrificing the second or third cuts if I have to. If you want a constant supply then you'll need to sown coriander every two or three weeks. Depending on the weather you will likely get two or perhaps three cuts from it before it bolts.

✚ Parsley is a little more robust, and is a biennial, though it is often grown as an annual since it is usually the leaf that we grow it for. There are a few types: flat leaf, curly and giant for instance. For an early crop of parsley try sowing in late summer, and planting out in early autumn. Curly varieties are a bit hardier and should cope fine in the ground over winter ready to give you fresh herbs in early spring as the temperatures start to increase. Such autumn planting also works well for growing the root to eat the following autumn.

How often to pick

Pick every one to three weeks.

ZERO-WASTE TIPS

✚ Coriander and parsley flowers are edible, with a delicate flavour that works well in salads.

✚ Coriander seeds are delicious. The traditional way to eat them is when fully mature and dry, either whole or ground. Try harvesting some seed while it is still green and fresh. It has a real zest to it; try in cheese sauce with pasta, or in stir-fries. Parsley seed is edible too, and can replace fennel seed in most recipes.

✚ Leave parsley plants for a second year to get a root crop from them. Hamburg parsley has been specially bred for its larger edible root.

TOO MANY?

✚ Fresh leafy herbs like coriander and parsley freeze well in butter or oil.

✚ You can make pesto from either of these herbs to use up big gluts.

✚ I often let them flower and seed rather than picking them all as leaf.

Try harvesting some coriander seed while it is still green and fresh. It has a real zest to it; try in cheese sauce with pasta, or in stir-fries.

MINT

Mentha spicata and others

Sow:	Early spring
Plant:	Spring
Harvest:	Spring to autumn
Eat:	Fresh, spring to autumn; frozen/ bottled, year-round

How often to pick

Pick as needed, but regularly picking ensures new young tender growth.

One of the easiest crops to grow is mint, which works well either in the ground or in containers. Pick often and, if growing in pots, divide and repot every year to keep it producing lots of new fresh leaf.

GROWING TIPS

✚ Though most mints grow from seed, I would buy one plant and grow on and divide it, unless you need lots of plants. Mint grows so vigorously that you will soon have all the mint you need.

Yield

	Per Plant	Per m^2/yd^2
Total	200g/7oz	5kg/11lb
Per pick	50g/2oz	1.25kg/2¾lb

SEEDS

Nigella damascena Nigella
Papaver somniferum Poppy
Helianthus annuus Sunflower
Foeniculum vulgare Fennel

While there are many plants whose seeds you can eat, here are some that are worth growing just for the seed. All of them are beautiful so you can plant them in ornamental areas of the garden or among your vegetables. Just remember to catch the seed before it is shed all over your garden.

Sow: Spring: nigella and poppy direct; sunflower and fennel in modules or pots

Plant: Late spring or early summer (sunflower and fennel)

Harvest: Late summer to early autumn

Eat: Fresh; or stored in dry airtight jar, year-round

Yield

	Per Plant	Per m²/yd²
Nigella	3g/0.1oz	120g/4oz
Poppy	25g/1oz	500g/1lb
Sunflower	100g/3½oz	600g/1¼lb
Fennel	10g/½oz	120g/4oz

GROWING TIPS

I eat a lot of seeds. It's where plants pack all the nutrition to feed their 'children'. As a seed germinates the baby plant first uses the stored food in the seed to send out its root to find nutrients and its leaves to find sunshine.

✚ Whether it's the crunch of poppy seed, the fragrance of nigella or the pungency of fennel, it's worth growing some plants for seed.

✚ Nigella is a stunning plant. My garden is full of self-seeded plants, so I don't consciously grow them any more, and just harvest a portion of the seed when I start to hear the seeds rattling in the pods each year, leaving the rest to give me next year's crop. To get them going, just scatter seed on some bare ground, give it a quick rake, and you will almost certainly get a crop.

✚ Establish poppies in the same way. They are a little more temperamental, although similarly once you have a good population in your garden it will self-seed readily. In the early stages keep any eye out for slugs, and weed if necessary.

✚ Sunflowers need more attention. They form big plants, particularly if you choose the larger-headed varieties that are good for seeds. You may need to stake the plants to prevent them blowing over. Birds also love the seed, so make sure you get them first. You can harvest the whole plant and hang it upside down in a dry place to finish the seed-ripening process.

✚ If you want a good quantity of fennel seed it's best not to leave it to a few chance bolted vegetable plants but instead to plant a patch deliberately for seed. I tend to plant perennial fennel as you can harvest its leaf as a herb and then get the seed too.

TOO MANY SEEDS?

✚ If you have too many poppy or sunflower seeds, you can add them to bread or toast lightly to add to salads.

✚ Nigella seeds are delicious in stir-fries and curries.

✚ If you really have too many to eat, scatter them in your garden for a crop of flowers.

Birds also love the seed, so make sure you get them first. You can harvest the whole plant and hang it upside down in a dry place to finish the seed-ripening process.

CURRANTS

Ribes rubrum

Sow:	n/a
Plant:	Late autumn or winter, while dormant
Harvest:	Summer
Eat:	Fresh, summer; frozen/preserved, year-round

These are some of the tastiest fruits and I love them straight from the bush. If you like fruity puddings or jam, then currants are a must, being mostly easy to grow and prolific croppers.

GROWING TIPS

✚ Blackcurrants are one of the easiest fruit crops to grow. Blackcurrants can grow up to 2m/6½ft in height and width, while white and red might reach 1.5m/5ft.

✚ For all bushes, regular pruning will make sure you keep getting good yields. As a general rule take out one-third of the growth each year.

✚ Birds love currants so netting is a good idea.

✚ Currants bushes crop well for 8–10 years and then become less productive so you will need to replace them. If your plants are healthy you can take hardwood cuttings from existing bushes to get new plants.

TOO MANY CURRANTS?

✚ Jam is a given but the richness of flavour in all currants makes them ideal for cordials or even wines and liquors.

✚ You can dry them to use in herbal teas, or flapjacks and biscuits. They freeze beautifully as well

Yield

	Per Plant	Per m^2/yd^2
Black currants	4.5kg/10lb	2.25kg/5lb
Red/White currants	3.6kg/8lb	1.75kg/4lb

MELON

Melo spp.

Sow:	Spring, in pots; direct in hotter climates
Plant:	Early summer
Harvest:	Summer to early autumn
Eat:	Fresh, summer to early autumn

Fresh, fully sun-ripened melons are out of this world. They demand rich soil, warmth and plenty of water. In cooler climates they need a greenhouse or tunnel.

GROWING TIPS

✚ Newer varieties are making it easier to grow melons in northern climates, but you should still look for varieties with shorter maturing times.

✚ Melons have a germination temperature of 25–32°C/77–90°F and a growing temperature of more than 21°C/70°F. If the night-time temperature regularly drops to less than 10°C/50°F the fruit will lose flavour.

✚ In very hot weather the plants may lose a few flowers. Provided it cools again the plants should recover.

✚ Melons like less-frequent but heavy watering. Give them a good soak once or twice a week.

ZERO-WASTE TIPS

✚ Melons don't store well, so eat or process quickly.

TOO MANY MELONS?

✚ Melon juice is delicious, as are melon ice lollies.

✚ To dry melon, slice and dry in a dehydrator or low oven.

Yield

	Per Plant	Per m²/yd²
Total	6kg/13lb	3kg/6½lb
Per pick	2kg/4½lb	1kg/2¼lb

RASPBERRY

Rubus idaeus

Sow:	n/a
Plant:	Winter
Harvest:	Summer or autumn – variety dependant
Eat:	Fresh, summer to autumn; preserved, year-round

Raspberry plants are very easy to grow. These plants work really well with chickens, which eat the pupae of the raspberry beetle, and also give a bit of fertility.

GROWING TIPS

✚ Raspberry plants don't form a bush but spread by underground runners which pop up. Assign a patch and let them fill it, rather than trying to keep them too confined.

✚ **Summer fruiting** – also called 'floricanes', which fruit on the previous years' canes.
Autumn fruiting – otherwise known as 'primocanes', which give fruit on the current year's canes.

✚ The major pest is the raspberry beetle. You'll know you've got them because parts of the fruit will look brown or black and shrivelled. Try hoeing the ground around the plants in spring and early summer as the larvae start to emerge to bring them to the surface for birds to eat.

ZERO-WASTE TIPS

✚ Raspberry leaves can be used to make tea (always check with a registered herbalist, particularly if you are pregnant).

TOO MANY RASPBERRIES?

✚ Dry the fruit to make tea or add to cereal.
✚ Raspberries freeze well or can be used for jams.

	Per Plant	Per m²/yd²
Total	500g/1lb	2.75kg/6lb
Per pick	90g/3oz	540g/1.2lb

BLUEBERRY

Vaccinium spp.

Sow:	n/a
Plant:	Autumn or winter, while dormant
Harvest:	Summer
Eat:	Fresh, summer; frozen/preserved, year-round

This fruit is great in pots if you don't have the acidic soil in your garden needed to grow them in the ground.

GROWING TIPS

✚ The most important thing to check before planting blueberries is your soil pH. They prefer an acidic pH5–5.5, though will cope up to pH6. If your soil is more alkaline than this, then use an acidic compost and grow these plants in pots.

✚ Water with rainwater, particularly if growing in pots and you live in a hard-water area. Hard water contains calcium, which makes the water alkaline and if regularly added to the compost will raise its pH.

✚ Prune blueberry bushes by cutting back about one-third of the branches to the base.

ZERO-WASTE TIPS

✚ Blueberry fruit doesn't mature all at once, so you'll need to keep an eye on the bush and pick regularly.

TOO MANY BLUEBERRIES?

✚ Can be added to any cake or dessert and freeze well.

Yield

	Per Plant	Per m^2/yd^2
Total	3–8kg/6½–17½lb	750g-2kg/1½–4½lb
Per pick	750g-2kg/1½–4½lb	200–500g/7–17oz

STRAWBERRY

Fragaria × ananassa

Strawberries are probably the perfect beginners' fruit, being easy to grow and almost guaranteed to get some fruit in the first year after planting. They do well in the soil or in containers, so even if you have only a balcony then try strawberries. I love the little alpine ones, though they are more fiddly to pick than the June-bearing varieties.

Sow:	Late winter (alpine varieties)
Plant:	Plant in spring or late summer (runners)
Harvest:	Summer
Eat:	Summer, fresh; preserved, year-round

Yield

	Per Plant	Per m^2/yd^2
Total	300g/10oz	1.75kg/4lb
Per pick	125g/2.5oz	300g/10oz

How often to pick

Pick every one to three days.
Fruit ripens more quickly in very hot weather.

Varieties

There are four main types of strawberry:

Tiny alpine varieties – these can be raised from seed, and planted out as ground cover around trees or bigger bushes, or in open borders. They will keep producing for quite a few years (though the fruit will tend to get smaller as they get older).

June-bearing – these give one big crop of fruit in early summer. They are good if you are looking to do lots of preserving, as the fruit all comes at once for easy picking.

Everbearing – the fruit in theory comes all through the summer, though in practice this usually means two main crops with a gap in the middle.

Day-neutral – an improved type of everbearing with three main crops through the summer.

GROWING TIPS

✚ You should be able to get at least six plants per m²/yd² with about 30cm/12in between plants, and rows 60cm/24in apart

✚ Strawberry fruit is very susceptible to rot, so try to keep it dry. Historically this would have been done by placing it on straw, which can work well. Composted woodchip also helps to keep the surface of the soil dry.

✚ Planting on ridges is another way to keep strawberries dry, and also has the advantage that the fruit hangs down the side of the ridge and is easier to see when picking. If you do plant in ridges, try leaving a little dip immediately around the plant; otherwise when you water you might find it all runs down the ridge and away from the plant.

ZERO-WASTE TIPS

✚ The stems and leaves of strawberries are perfectly edible, and though not that tasty in themselves if you are juicing or making smoothies then you can just leave the stalk on. This saves time prepping and wastes less.

✚ Strawberry plants send out runners, which will root into the soil around if they can. To keep your main plant productive you need to cut these off. Rather than throw them all on the compost heap, allow some to root first and then pot them up for new free plants.

TOO MANY STRAWBERRIES?

✚ Strawberries freeze perfectly adequately, though like most soft fruit they come out mushy. This is fine for anything where they don't need to look good.

✚ Ice creams and sorbets are great, but take time and freezer space. One of my favourite quick desserts is to mash up a pile of fresh strawberries and mix in some Greek yogurt – even better when sprinkled with a few nuts or seeds.

✚ Try drying strawberries for a more intense-flavoured snack.

APPLE & PEAR

Malus domestica, Pyrus spp.

Apples and pears are some of the easier tree fruits to grow, and they cope with a range of climates and situations. Small trees can do well in pots if you don't have much space, or you can train them against a wall or fence to make the most of your garden space.

Sow:	n/a
Plant:	Winter, while dormant (bare-root trees); year-round (pot-grown trees)
Harvest:	Late summer to winter
Eat:	Fresh, late summer to winter; stored, winter to early spring; preserved, year-round

Yield

	Per Plant	Per m²/yd²
Small tree e.g. 'M9' apple or 'Quince C' pear rootstock	11–18kg/ 24–40lb	1.2–2kg/ 2–4½lb
Large tree e.g. 'M25' apple or 'Communis' pear rootstock	90kg/200lb or more	1.75kg/ 4lb

GROWING TIPS

✚ Though there are size differences between varieties, the final dimensions of your tree is mostly determined by the rootstock it has been grafted on to. For pots or small gardens look for dwarfing rootstocks. If you have enough space, a more vigorous rootstock will give you a larger stronger tree.

✚ Soil quality also impacts tree size. On poor soils plant a stronger rootstock to compensate, while on very fertile soils a weaker rootstock will still do well.

✚ More vigorous rootstocks also give better-quality fruit. My fruit guru, Matthew Wilson, a wonderful grower from Sussex in England, once told me that he sells the fruit from the smaller weaker rootstocks first and stores that from vigorous rootstock as they keep better.

✚ Apples and pears need trees of a different variety to pollinate them. For most of us there are enough other trees in neighbours' gardens or wild crab apples in the hedgerows, but if you are on your own and kilometres from the nearest apple tree then check pollination groups before you buy your trees.

✚ I won't presume to give too much advice on varieties – there are thought to be more than 7,000 apple varieties worldwide for instance – except that I would opt for a reliable cropper that does not rely on feeding or pest and disease treatments. Avoid some of the very old varieties; they are often heritage crops for very good reasons and may give a disappointing yield. At the other end of the scale, some of the very modern commercial varieties have been bred for high-input systems and don't always perform so well in the garden.

✚ Fitting everything you want into your garden is always an issue. Apples and pears can easily be trained into almost any shape to make the best use of space. Try them as a single 'stepover' trees running along bed edges or as fans and espaliers against trellis, fencing or walls.

✚ Don't be put off by the pruning, which is easily learnt.

TOO MANY APPLES AND PEARS?

✚ Small gluts can be used for crumbles, chutneys or in cakes. Sliced apple rings are also wonderful when dried.

✚ Once trees really start producing you will probably not keep up. The only way then is to start juicing. As a general rule for every 1kg/2lb of fruit you should get 0.5 litre/17fl oz of juice. You can drink it straight away or store it in the fridge for a few days. To keep the juice longer you'll need to pasteurize it. This can be done in a saucepan for small quantities.

✚ Small-scale juicers are readily available, or in some areas there are community juicing events where you can bring your apples along and get them juiced. Some juicing companies will also juice and bottle your harvest. If you have enough (for instance 250kg/550lb) they will juice as an individual batch. They might also offer a service where smaller quantities can be added together with other batches and you get some of the mixed juice.

✚ You might also like to start making your own cider (Hard cider in the USA). Not only will this use lots of apples, but you will also be able to use the apples that have been bruised or damaged.

PLUM

Prunus spp.

Sow:	n/a
Plant:	Winter, while dormant (bare-root trees;) year-round (pot grown)
Harvest:	Summer
Eat:	Fresh, summer; preserved, year-round

Plums are a good beginners' fruit. They need little attention and are one of the most colourful fruits

GROWING TIPS

✚ Water them well in the first year; a good soak every couple of weeks and an organic mulch should be fine.

✚ The other big risk with plums is that the branches can get so laden with fruit that they break. There is no strict rule but you might need to remove one-third of the fruit in very heavy years. As a guide leave about 7cm/3in between fruits.

✚ Prune in late spring or early summer. Most will need little pruning once established.

✚ Be a bit careful when picking as wasps love plums.

ZERO-WASTE TIPS

✚ To best use your space try growing cherry plum (*Prunus cerasifera*). It forms a great dense hedge but also produces delicious small plums.

TOO MANY PLUMS?

✚ Plums bottle really well, or you can dry them into prunes.

✚ You can also make plums into a wide range of drinks.

Yield

	Per Plant	Per m²/yd²
Small tree – e.g. 'Pixy' rootstock	10–15kg/ 22–33lb	2–3kg/ 4½–6½lb
Large tree – e.g. 'St Julien A' rootstock	18–25kg/ 22–33lb	2–2.75kg/ 4½–6lb

GRAPE

Vitis vinifera

Sow:	n/a
Plant:	Spring
Harvest:	Summer to early autumn
Eat:	Fresh, summer to early autumn; preserved, year-round

Grapes hold a special place in my heart. With the supermarkets now so full of hard tasteless bunches it is a joy to taste the full flavour of home grown grapes.

GROWING TIPS

✚ You'll need to decide whether you are growing for wine or eating before choosing which variety. In cooler climates it is safer to go with a white variety that will give a decent crop without long hot summers, or try growing in greenhouses or tunnels. Seeded table grape varieties tend to be the hardiest.

✚ Grapes are climbing plants and need support. Vines are flexible to form; you will be training back to a permanent framework of branches but what form that framework takes is up to you.

✚ Thining out your grape bunches can produce better fruit.

ZERO-WASTE TIPS

✚ Though we normally eat only the fruit, vine leaves are perfectly edible. They can be stuffed with fillings, added to stir-fries or omelettes instead of spinach.

TOO MANY GRAPES

✚ You can make wine, but fresh grape juice is delicious or dry the grapes to make your own raisins or sultanas.

Yield

	Per Plant	Per m²/yd²
Total	5–10kg/11–22lb	1.6–3.25kg/3½–7lbs

INDEX

GLOSSARY

Bare root – used to describe a transplant or tree that has been grown in the ground and then pulled up (the soil comes off the root, hence the term "bare root") and then transplanted. Fruit trees usually establish better when bare root as do some vegetables like cabbages and leeks.

Biochar – a form of charcoal (biomass burnt in controlled conditions in the absence of air). Has potential to improve soil health in some circumstances, sequester carbon and hold onto water and nutrients in soil and composts.

Blanch (cooking) – putting in boiling water briefly to kill bugs that might spoil the produce.

Blanch (gardening) – to keep light away from a growing plant in order to get soft delicate growth. Can be achieved by wrapping paper around celery stalks or growing rhubarb in the dark, for example.

Bolt – when a plant flowers and goes to seed, often prematurely, usually because of hot and/or dry weather.

Cloche – a transparent cover used to protect plants from cold and wind, like a mini green house or tunnel.

Damping off – a disease that affects young seedlings, usually because they have been sown too thickly and/or watered too much

Dibber – a tool designed for making holes in the ground for planting

F1 hybrid seed – F1 stands for first generation and these varieties are the result of a controlled cross between 2 known parents. In perfect growing conditions they will usually give a more vigorous plant and greater yield, and so are often used by commercial growers. However, being genetically similar they are less resilient if grown in less favourable conditions.

Grafting – the joining of two different plants (usually the roots of one and a stem of another) to make one "grafted" plant.

Hardening off – the process of toughening plants that have been grown in greenhouses of tunnels before planting out. Usually involves bringing them outside during the day and in again at night for a week or so.

How long do plants live?

- **Annual** – grow, seed and die in one year, for example, lettuce, sunflower.
- **Biennial** – grow in first years, seed and die in second year, for example, carrot, beetroot
- **Perennial** – live for a number of years producing seed each year, for example, asparagus, blackcurrant

Leaves

- **True leaves** – the main leaves of the plant that develop after the seed leaves. Often you will see descriptions that say "plant out when three or four true leaves have developed".
- **Seed leaves** – when a plant first geminates it will send up one or two leaves (depending on plant type) these are usually different to the true leaves that develop afterwards.

Ley – a temporary grass or mixed cover planting between crops.

Modules / cells – trays of very small pots, great for saving space when propagating plants.

Multisown – when a few seeds are sown together into a module or pot.

Open pollinated seed / variety – these are varieties that when you save the seed will give a plant of the same variety. They have a degree of genetic diversity that makes them more adaptable to a range of conditions

Prick out – transplanting seedlings into modules or pots. This needs to be done carefully to make sure the stems and roots are not damaged. Hold seedlings by the leaf as if the leaf breaks it will regrow; if the stem is damaged the seedling will probably die.

Rootbound – when the roots of a plant get too big for the module or pot it is in. Often they start to grow in circles around the inside of the plant, resulting in plant stress and less chance of success once planted out.

Seed drill – a shallow groove or trench into which seed is sown, and then covered with a thin layer of the soil.

Seedbed – soil prepared for sowing seed

Soil pH – Term given to describe the acidity/alkalinity of soil 7 is neutral. Acidic soil goes down numerically and alkaline soil goes up. The scale is exponential so six is ten times as acidic as seven and five is ten times again. Very easy to test and important to know when growing certain crops like blueberries.

Tilth –the structure of soil in a seedbed. A fine tilth (needed for sowing crops like carrots) would be like sand.